Ecology

Principles and practice

W H Dowdeswell, M.A., F.I.Biol.

Emeritus Professor of Education, University of Bath

 Heinemann Educational Books

Heinemann Educational Books Ltd
22 Bedford Square, London WC1B 3HH

LONDON   EDINBURGH   MELBOURNE   AUCKLAND
HONG KONG   SINGAPORE   KUALA LUMPUR   NEW DELHI
IBADAN   NAIROBI   JOHANNESBURG
EXETER (NH)   KINGSTON   PORT OF SPAIN

First published 1984

British Library Cataloguing in Publication Data

Dowdeswell, W. H.
  Ecology.
  1. Ecology
  I. Title
  574.5   QH541

  ISBN 0-435-60226-8

Cover and text illustration by Sam Denley
Typeset by Eta Services (Typesetters) Ltd, Beccles, Suffolk
Printed in Great Britain by Fletcher and Son Ltd, Norwich

Contents

Preface   v

Introduction   vii

1   Ecology in context   1
2   Patterns and distribution   14
3   The dynamics of populations   28
4   Communities and ecosystems   49
5   Methods of study: Collecting and sampling   72
6   Methods of study: Measuring the environment   98
7   Methods of study: Using information   127
8   Ecological communities: Soils and woods   149
9   Ecological communities: Downs and dunes   166
10   Ecological communities: Ponds and streams   181
11   Ecological communities: Sea shores and estuaries   210
12   Ecological communities: Heaths and bogs   236
13   Ecological communities: Man-made habitats   249
14   Ecology and man   268

Bibliography   294
Some useful addresses   299
Glossary   302
Index   308

Preface

Ecology : principles and practice is intended as a reader and source of reference for school sixth form and first year university students. As it is orientated towards problems of studying the subject, it is hoped that it will also provide a useful guide for teachers. The book divides roughly into four parts. The general principles of the subject occupy the first four chapters. These are followed by a section on methods of study: the collection, handling, and interpretation of data (Chapters 5–7). The application of ecological principles, methods for the study of some of the commoner habitats in Britain, and some suggested problems for investigation occupy the next six chapters. The final chapter is on ecology and man and includes the problems of pollution and conservation.

Some years ago I wrote two books on ecology which achieved considerable success in their day but are now out of print: *An Introduction to Animal Ecology* (Methuen) and *Practical Animal Ecology* (Methuen). In preparing the present volume, I have made use of small amounts of material from these publications that have stood the test of time, particularly concerning ecological techniques and the handling of data. However, it is now accepted that at a more advanced level, plant and animal ecology need to be treated together as a single entity, and it is this approach that has been adopted here. Great advances have occurred recently in fields such as the electronic measurement of physical factors and the use of computers in handling data and simulating ecological situations. Such developments are taken fully into account in this book.

A list of references is included at the end of each chapter, also a classified bibliography at the end of the book. A catalogue of useful addresses has been compiled as an aid to those wishing to pursue particular aspects of ecology beyond the coverage in this book. Technical terms which are likely to be unfamiliar to students are defined as they arise and summarised in the glossary.

For valuable help and advice during the writing of this book I am in debt to a wide circle of friends. Professor E. B. Ford, FRS, kindly read the early and more general chapters, and I am grateful to him for his penetrating and helpful comments. John Barker read the whole of the draft typescript and my thanks are due to him for the care he took and for his many helpful suggestions. Graham Taylor's help and comments have also been most valuable, as has the editorial work of Penny Hayes. The illustrations are derived from many sources, but in this connection I would particularly like to thank Alison Leadley Brown for the beautiful drawings of freshwater plants and animals used in figures 10.4, 10.10, 10.18, and 10.19. The sources of the other figures are acknowledged individually and my thanks are due to those authors who have kindly permitted me to make use of illustrations and data from their publications.

Photographs play an important part in the illustration of this book. The majority are my own, and in this connection I would like to thank Tony Wheeler of Bath University for making excellent prints from many difficult negatives, including conversion from colour to black and white. My thanks are also due to those who have supplied me with photographs, some of them taken specially for the purpose. Mr S. Beaufoy provided figures 3.7(b), 3.11, 10.5, 12.8, 13.7, 14.7; Professor A. D. Bradshaw, figure 2.7; John Clegg, figures 10.6, 10.9 and 10.13; John Haywood, figure 14.4; Eric and David Hosking, figures 11.3, 11.5, 11.7, 11.10 (on behalf of Dr D. P. Wilson), 12.6; Frank W. Lane Agency figures 1.5, 3.7(a), 4.13, 13.14; Dr H. Canter-Lund, figure 10.12; Philip Harris Ltd, figures 5.3, 5.4, 5.5, 6.1, 6.4, 6.6, 6.14, 6.15; Griffin and George Ltd, figures 5.10, 5.11, 5.12, 5.15, 6.7, 6.12; Turner International Engineering Ltd, figure 13.3; The Forestry Commission, figures 3.6, 14.5 and 14.6.

In writing this book I have been greatly assisted by Elaine Cromwell and I would like to thank her for her patience in deciphering my writing and for her unfailing efficiency in the preparation of the typescript and the index.

W. H. Dowdeswell
1983

Introduction

In recent years much has been said and written about the science curriculum at all levels in our education system. The study of biology, in some form or other, occupies an important place in this curriculum, if only because, being concerned with all aspects of life, it impinges closely on man himself. One of the greatest problems for students today is that we live in a world where the fund of knowledge is still increasing exponentially. There is just not time to learn all that we would wish, so the problem arises as to what should be left out.

There are those who contend that topics such as ecology occupy only the fringe of biology and therefore can be omitted even from first year university courses in favour of such subjects as anatomy, physiology, and genetics. After reading this book, I can only hope that the student will have been persuaded otherwise.

The outstanding claim of ecology as a branch of study is that it is concerned with living things as they really are, occupying a diversity of places and responding to one another and their physical environment in a variety of complex ways. If this book has a single aim above all others, it is to explore these responses and to suggest methods of studying them. Much of biology learning today is still wedded, at least in part, to the approach advocated by T. H. Huxley late last century. This involved the selection of a few type species to illustrate different levels of biological organisation. Plants and animals do not, however, exist in isolation and it is the purpose of ecology to clarify and explain the nature of the interactions on which wild populations and communities depend. As such, the subject must surely qualify as one of the corner stones on which the edifice of biology rests.

However, there is more to ecology than just its academic aspects. Part of the excitement of the subject derives from the fact that it brings us face to face with some of the greatest problems of our time. In a small island such as ours, the rival claims of such interests as housing, agriculture, industry, and amenity pose major problems in the partitioning and maintenance of a limited environment. Recently, there has been a marked increase in public awareness of the need for conservation in all its aspects, both for the survival of wild populations of plants and animals and for the needs of human society. A number of factors have combined to bring about this change of attitude, including programmes on television, the publicity given to problems of pollution, and greatly increased access to the countryside. All of these have ecological implications.

There is, therefore, no need to justify a book on the study of ecology which covers modern approaches and techniques, and the broader aspects of ecology which are of contemporary importance. The argument for the inclusion of such topics in courses of biology at all levels, both for their academic worth and their practical significance, is overwhelming.

The author and publisher would like to acknowledge the following sources from which tables were taken and figures were redrawn with permission:

Fig. 1.4, Schlieper, P. (1952), *Biol. Centralbl.* **71**, 452; Fig. 2.3, Reynoldson (1961), *Verh. int. Ver. Limnol.* **14**, 989; Fig. 2.4, Wadsworth R. M. Ed. (1968), *The Measurement of Environmental Factors in Terrestrial Ecology*, Blackwell Scientific Publications Ltd; Fig. 2.5, Macan, T. T. and Worthington, E. B. (1951), *Life in Lakes and Rivers*, Collins; Table 2.1, Edington, J. M. (1968), *Journal of Animal Ecology*, **37**, 688, Blackwell Scientific Publications Ltd; Fig. 2.9, McNeilly, T. S. (1968), *Heredity* **23**, 103; Fig. 3.3, *Ecology* 2nd Ed. by Eugene P. Odum. Copyright © 1975 by Holt, Rinehart and Winston. Reprinted by permission of Holt, Rinehart and Winston, CBS College of Publishing; Fig. 3.8, Ford, E. B. (4th edn, 1975), *Ecological Genetics*, Chapman and Hall; Table 4.1, Neal, E. G. (1948), *The Badger*, Collins; Fig. 4.5, The Open University, Block A S323, *Energy Flow Through Ecosystems 5 Whole Ecosystems*, © 1974 The Open University Press; Fig. 4.7, King, T. J. (1980), *Ecology*, Nelson; Fig. 4.10, *Ecology* 2nd Ed. by Eugene P. Odum. Copyright © 1975 by Holt, Rinehart and Winston. Reprinted by permission of Holt, Rinehart and Winston, CBS College Publishing; Fig. 4.12, Hardy, A. C. (1955), *The Open Sea, The World of Plankton*, Collins New Naturalist; Figs. 5.17, 5.19, and 8.8, Ashby, M. (2nd edn. 1969) *An Introduction to Plant Ecology*, by permission of Macmillan, London and Basingstoke; Fig. 5.18, Fry, P, Ed. (1971) *Nuffield Advanced Biology, Laboratory Book*, © The Nuffield Foundation, published by Longman Group Ltd 1974; Table 8.1, Edwards, C. A. and Lofty, J. R. (1977), *The Biology of Earthworms*, Chapman & Hall Ltd; Fig. 6.16, Former BA Chart No. C6104, with the sanction of the Controller H.M. Stationery Office and of the Hydrographer of the Navy; Fig. 8.11, Darlington, A. (1972) *The World of a Tree*, Faber; Table 10.3, Varley, M. E. (1967) *British Freshwater Fishes*, Fishing News Books, London; Fig. 10.3, Mills, D. H. (1972) *An Introduction to Freshwater Ecology*, Oliver and Boyd; Table 11.4, Nicol, J. A. C. (1961), *The Biology of Marine Animals*, by permission of Pitman Books Ltd, London; Fig. 11.4, Southward, A. J. (1965) *Life on the Sea Shore*, Heinemann; Fig. 11.18, Humphreys, T. J. (1981) 'Estuarine ecology as project work', *Journal of Biological Education*, **15**, 231; Table 12.1, Pearsall, W. H. (1950) *Mountains and Moorlands*, Collins; Figs. 13.5 and 13.8, Darlington, A. (1981), *Ecology of Walls*, Heinemann; Fig. 13.13, Darlington, A. (1969), *Ecology of Refuse Tips*, Heinemann; Table 14.1 and Fig. 14.2, Richardson D. H. S. (1975), *The Vanishing Lichens*, David and Charles; Fig. 14.9, Tribe, M. A. *et al.* (1974), *Basic Biology Course, 4 Ecological Principles*, Cambridge University Press; Fig. 14.10, BP Nutrition Ltd (1975), *A Protein Survey*, B.P. Nutrition; Table 14.4, Wilson, P. N. (1973), *Pnl. Trans. R. Soc. Lond. B.*, **267**, 101–2; Table 14.5, with permission from Perring, F. H. and Mellanby, K. eds (1977), *Ecological Effects of Pesticides*, Academic Press. Copyright Linnean Society of London; Fig. 14.12, I. F. Spellerberg, (1981), *Ecological Evaluation for Conservation*, Studies in Biology Series, Edward Arnold

1

Ecology in context

Everyone knows, or thinks they know, what ecology is all about. Indeed, the word is now so commonplace in our everyday vocabulary that it has even found its way into the field of politics! However, widespread use of a term does not necessarily imply an equally widespread understanding of what it means. The very obscurity of the word itself gives us little assistance. (Coined by the German biologist Haeckel in 1878 from two Greek words meaning 'house' and 'discourse'.) Why Haeckel should have felt that a new scientific term denoting the study of the interactions of plants and animals with one another and their environment was needed at that time is uncertain. Perhaps it was a reaction against the contemporary emphasis in both botany and zoology (biology as a subject barely existed then) on the study of the individual organism in isolation. To some extent that emphasis still exists today, particularly in secondary education.

The study of ecology

There are many arguments in favour of the inclusion of ecology in biology courses at all levels. It would be out of place here to rehearse all of these, but two of them deserve special mention. The first derives from the obvious fact that plants and animals do not exist alone. As Haeckel implied, they live together in **populations** (of the same species) and **communities** (of different species), interacting with one another and their physical surroundings in an intimate and usually complex manner. The second reason was highlighted by the distinguished British ecologist Charles Elton more than half a century ago. In the Preface to his book, *Animal Ecology* [1.1], written in 1926, he says, 'Ecology is a branch of zoology which is perhaps more able to offer immediate practical help to mankind than any of the others, and in the present parlous state of civilisation it would seem particularly important to include it in the training of young zoologists'. How perceptive Elton was! The only way in which we would wish to modify his statement today would be by substituting 'biology' for 'zoology'. We will return in later chapters to a consideration of the usefulness of ecology to man and its modern applications in the field of biotechnology and elsewhere.

Another reason why the study of ecology is important is that it helps us to relate one area of biology to another. As is illustrated in Fig. 1.1, it lies at the heart of the subject, having links with almost every other branch. Thus competition between members of a species (**intraspecific** competition) leads to variation through genetics and differential survival, and hence to evolution. All adaptations of living things to their environment are ultimately physiological and behaviour patterns also play an important role in survival. Again, the idea of the species is fundamental to ecological thinking and this relates to such fields as taxonomy and morphology. From the viewpoint of economics, a particularly important aspect of ecology is the soil and the factors connected with it (**edaphic factors**) which influence its populations and communities. This aspect of ecology links closely with certain areas of geography and geology. Marine ecology plays an important

1

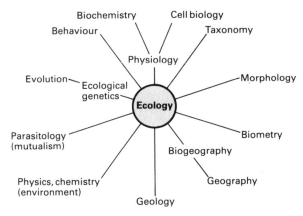

Fig. 1.1 Some relationships of ecology

part in furthering the success of our fisheries. Parasitism and other intimate relationships among plants and animals (sometimes referred to as **symbiosis,** see Chapter 3) together constitute a special branch of ecology with important implications for agriculture, for instance in the control of pests and disease. Finally, aspects of physics and chemistry permeate in varying degrees all facets of ecological studies, particularly those concerned with the physical (**abiotic**) environment and the ability of plants and animals to survive within it.

As the need to adjust to varying environmental conditions can bring about profound changes in the nature of populations and communities, so too, the organisms themselves can modify the environment in a variety of ways. Thus a common feature of trout streams is their colonisation by a range of aquatic plants such as the water crowfoot, *Ranunculus aquatilis,* which support the many small arthropods that provide the food of fish. However, when present in large masses, the roots of these plants are an effective barrier to the movement of silt washed along by the current. In a surprisingly short time the silt can accumulate to the point where an island is formed above water level, which will soon be colonised by a community of land plants. The job of preventing such changes occurring is one which habitually faces water bailiffs who are responsible for maintaining waterways in a condition favourable to the maximum production of fish. Incidentally, this example also illustrates the important general principle that **conservation** is an active process, in contrast to preservation which is essentially passive.

Of all the ecological factors capable of promoting change, man is unquestionably the most powerful. Indeed, it is probably true to say that there is now no portion of the earth's surface that has not been modified by human influence to some extent. Alas, many of these modifications have been for the worse, pollution and the depletion of resources being obvious examples (a matter to which we will return in Chapter 14). All living communities are governed by the simple laws of economics; thus consumers cannot exist in the absence of producers and for effective survival a balance must be struck between them. Human societies are subject to the same laws, so a study of the ecology of wild populations may well prove helpful in interpreting the ecology of man.

As we saw earlier, plants and animals living together in communities interact with one another. Thus in an area of grassland we could find butterfly larvae feeding on grasses, shrews eating the larvae and owls preying upon the shrews. We thus have the sequence:

soil – grass – larvae – shrews – owls.

Such a sequence of relationships is known as a **food chain**.

A situation of this kind where a number of distinct but related entities are linked together and mutually dependent is known as a **system**. In ecology the bonds existing between different organisms represent, among other things, the transfer of essential food materials, the whole being appropriately called an **ecosystem**. In general, there are two different ways in which we can approach the study of ecosystems. We can adopt a procedure in which we isolate individual species and study their ecological characteristics in some detail (**reductionist approach**). Using the example already quoted, we might begin with an investigation of grass tufts, the life cycles of the species making up the tufts and of the animals using them as food. Information on browsing herbivores could lead to the next links in the chain and a gradual build-up (**synthesis**) of some of the economic relationships of the community. In the early days of ecology, great stress was laid on conducting ecological surveys whose principal purpose was to provide a compendium of species and the generalised ecological backcloth against which they existed. The alternative approach (**holist**) is to make a study of the whole community in general terms first and then to break it down by the process of **analysis** into its component parts.

If we were to regard, say, an area of grassland as representing an ecosystem, then we may be sure that, in fact, it consists of a number of subsystems related to one another with varying degrees of intimacy. Availability of time, if nothing else, will usually restrict attention to only a few of these subsystems, and if we wish to obtain a broader view, such as of overall production and consumption in a particular area, it will be necessary to adopt a more crude approach. Such problems will be considered further in Chapter 4. Meanwhile, it is worth noting that **holism** and **reductionism** represent poles of a spectrum of enquiry. Most ecological studies involve an element of both. Bearing in mind that the full range of occupants of any habitat is what frequently faces us at the beginning of any ecological study, a holist approach followed by gradual analysis has much to recommend it. But no matter what approach we adopt, the starting point in any ecological investigation is with the smallest unit (**species**), which is the basis of identification. Defining a species is not easy and in order to apply the concept in the context of ecology it is important to appreciate the limitations of the various criteria generally employed. These are discussed further in the next section.

What is a species?

The concept of the species is often thought of as a means of providing a **systematic** (classificatory) method of identifying and naming particular plants and animals. However, this is rather a narrow view, for its significance pervades many kindred areas of biology, not least that of ecology. Crowson [1.2] and other writers have reviewed the criteria used in defining species as basic units of taxonomy. These can be summarised briefly as follows.

(i) Morphological criteria

The structure and appearance of an organism, such as its colour and adult size, are the characteristics most frequently used for the identification of plants and animals in museums and in the field. Difficulties in applying these criteria arise from the fact that all organisms are subject to different amounts of variation, both genetic and environmental. So if speciation is to have any meaning, some degree of discontinuity must occur between one species and another closely resembling it.

(ii) Physiological criteria

Just as certain morphological features are characteristic of a species so, too, are certain physiological and biochemical factors. For instance, the range of enzymes present in blood and other body fluids can provide an important index of variation in both vertebrates and invertebrates, which can readily be assessed by such methods as electrophoresis.

(iii) Genetic criteria

Matings between individuals of the same species normally produce fertile offspring while crosses between different species are usually unsuccessful. When crosses between different species do occur, the offspring are generally infertile, but not invariably so. Thus, while the cross between a horse, *Equus caballus*, and a donkey, *E. asinus*, produces a mule, which is usually sterile, fertile mules also exist as rarities. The bladder campion, *Silene vulgaris*, is a tall, erect species and a common colonist of hedgerows, while the sea campion, *S. maritima*, is an inhabitant of cliffs and shingle. Where their habitats overlap they interbreed freely, but the hybrids are adapted to neither situation and so remain localised and relatively rare.

(iv) Geographical criteria

A species normally occupies a definite geographical range, being separated from kindred species by a distinct discontinuity. This is frequently due to some ecological barrier, such as a mountain range, a stretch of water or an area of inimical land such as a desert. However, areas of overlap between two related species do occur, leading to coexistence, each population retaining its identity and existing separately from the other. Sometimes a degree of hybridisation may occur, as between our two native species of crow. Thus the carrion crow, *Corvus corone*, is restricted in Europe to western and south-western areas, while the hooded crow, *C. cornix*, has a wide distribution to the east and north. The birds are powerful fliers and stretches of water such as the North Sea are not barriers to their movement. When their geographical ranges overlap, for instance in north Scotland, they form hybrids, some of which are fertile. However, relative to the total range of the two species, the area of mixing is quite small.

(v) Ecological criteria

Members of the same species tend to exhibit similar behaviour patterns and habitat preferences. Hence, they occupy a particular economic position in any community which is not usually shared with any other species.

(vi) Palaeontological criteria

A species occupies a distinct and limited range in geological time. However, the application of palaeontological criteria has limitations for two main reasons. First, as Darwin pointed out, the fossil record is incomplete and seldom, if ever, is it possible to follow through the fossil history of a species in its entirety. A second limitation is that while fossils can provide evidence relating to time and spatial distribution, they give only a limited idea of the ecological conditions that prevailed during a particular era.

The application of species criteria

The simultaneous application of all these criteria is clearly impossible and usually the identification of an individual plant or animal will depend mainly upon morphology and to a lesser extent, on geographical and ecological characteristics. Here again, difficulties may arise. For instance, many characteristics which are useful for identification, such as colour and size, are under polygenic control and therefore exhibit a range of variation from one extreme form to the other, e.g. large to small, or dark colour to light. Sometimes these variants may exhibit geographic gradients with a high value at one extremity of distribution and a low value at the other. Such gradients of variation are known as **clines** and their extent may vary from a few hundred yards of hedgerow to several hundred miles of diverse country. Specimens reaching museums for identification are frequently derived from the latter situation, having been collected at only a few points along the cline and hence being far from representative of the species as a whole.

The occurrence of **polymorphism** may also cause difficulties, as the species can then occur in two or more forms each of which is under genetic control. A typical example among plants is the existence of pin and thrum flowers in the primrose, *Primula vulgaris* (Fig. 1.2 p. 6 and Fig. 7.7 p. 139). This polymorphism promotes cross-fertilization. Polymorphism among animals is also widespread, for instance the black (melanic) and light forms of many moths such as the peppered moth, *Biston betularia* (Fig. 14.4 p. 272) and the human blood groups.

Compared with the views of the great systematists of the eighteenth century, such as John Ray and Karl Linnaeus, to whom we are indebted for founding the modern methods of taxonomy, our present views of species have greatly changed in the light of an increased knowledge of the extent of variation in plants and animals, and of how it is controlled. Among botanists there has been a tendency to accommodate the wide range of variation observed among plant species by having recourse to ever smaller groupings, particularly sub-species. Exponents of this approach have been referred to colloquially as 'splitters'. Zoologists on the other hand have tended to regard the species as including a fairly wide range of variation resulting in what Mayr [1.3] has called the **polytypic** species. In contrast to botanical practice, the zoologists are sometimes, therefore, referred to as 'lumpers'.

Principles and processes

In spite of the limitations outlined above, the fact remains that the species concept has served ecology well and presented relatively few difficulties in practice. The binomial system of nomenclature using both generic and specific names, which

(a)

(b)

Fig. 1.2 (a) Pin and (b) thrum flowers of the primrose, *Primula vulgaris*. (See also Fig. 7.7, p. 139)

was introduced and developed by Linnaeus in the eighteenth century, is now employed universally for the identification of plant and animal populations. As we have seen already, due to their inherent variation and interactions with their environment, such populations are in a constant state of change. The concept of species is therefore dynamic, and if it is to serve the purpose of the ecologist it must remain so, no matter what system of categorisation is employed.

Ecology is a complicated subject in that it is concerned not only with change but with the great variety of interactions within populations and communities that bring it about. In any account, it is inevitable that a reductionist approach is adopted to some extent, so that different ecological components can be treated as separate entities. This is the procedure that will be followed in subsequent chapters. But before proceeding to such an analysis it may be helpful to view ecology from a holist standpoint and to attempt to identify underlying processes and ways in which they are related.

The process of colonisation

The colonisation of a bare patch of ground is to some extent a matter of luck depending on the chance arrival of **propagules** (agents of propagation), usually the seeds of flowering plants. As primary colonists, the common weeds such as groundsel, *Senecio vulgaris*, are particularly successful, as any gardener knows. Their reproductive capacity is high, as they produce large numbers of small seeds which are easily dispersed. However, food reserves in the seed are small, so unless a seed happens to land in a place where it can germinate quickly, it will perish. Larger seeds such as those of the Paplionaceae (vetches, clovers, and trefoils) are able to rely on their own resources for longer, the seedlings becoming more firmly established before they attain nutritional independence. For the effective establishment of any population of plants (or animals) appropriate conditions must prevail for reproduction and dispersal, there must be an adequate food supply, and the physical environment (including shelter) must be such as to promote survival and growth.

Sometimes, particularly in plants, a single physical (**abiotic**) factor can exert a predominant influence on the distribution of a species. As might be expected, light (the form in which energy is obtained from the environment) often acts in this way. The light intensity at which energy intake (**photosynthesis**) is just balanced by consumption (**respiration**) is known as the **compensation point**. This varies greatly from one species of plant to another. A typical shade plant not only has a lower compensation point than a sun plant, but also cannot make such good use of higher light intensities. However, the effect of light on plants can be more subtle than just determining whether a species can colonise a particular habitat or not. The rosebay willowherb, *Epilobium angustifolium* (Fig. 1.3), sometimes known as fireweed because of its capacity for colonising burnt ground, is a typical sun plant which is nonetheless frequently to be found in the shade of woodland. In this unaccustomed habitat, reduced photosynthetic activity is sufficient to support vegetative growth but insufficient to promote flowering. Such colonisation can, therefore, only be of a temporary nature, since environmental conditions prohibit sexual reproduction.

The influence of particular physical factors is frequently less clear-cut than in the example just quoted. More often (particularly in animals), two or more combine to exert a joint effect, one modifying the action of the other. The principle is well illustrated in fish such as the trout, *Salmo trutta*, where oxygen requirements are related to the temperature of the water and hence to the animal's rate of metabolism (Fig. 1.4). Thus, during the summer, trout flourish in rapidly flowing streams where oxygenation of the water is high and temperature fluctuations small, but they are far less happy in slowly flowing rivers where temperature changes may be considerable and oxygen concentration low.

Just as plant and animal populations must adjust to the abiotic environment in which they live, they must also adapt to living with other organisms of the same or different species. Within a population, competition will occur between individuals

Fig. 1.3 Rosebay willowherb, *Epilobium angustifolium*, a typical sun plant which colonises open ground

(**intraspecific competition**) for such commodities as food, mates, and living space. Within communities, competition between populations (**interspecific competition**) will take place with respect to such things as the range of habitats, production and consumption, predators and prey, parasites and hosts, and so forth.

Intraspecific competition is particularly fierce between males and between females (mating pressures) and where the availability of habitats (related to population numbers) and a sufficiency of food are concerned. Such factors are said to be **density-dependent**. Among plants, population density will depend largely on the capacity of the species to make use of the limited amounts of mineral

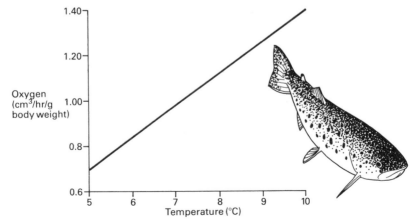

Fig. 1.4 Relationship between oxygen consumption and temperature in the trout, *Salmo trutta*

nutrients and light available in a particular locality. However, many animal species have evolved elaborate schemes for eking out a restricted food supply and regulating the density of the population depending upon it. Thus we find that the territorial system, involving the establishment of individual breeding areas, has developed to varying degrees among such diverse groups as butterflies, fishes, birds, and mammals. Its significance appears to be not only as a means of regulating resources but also in enhancing the chance of reproductive success.

The significance of variation

An important development in the study of ecology in recent years has been in the field of ecological genetics. Much of this work has centred on an assessment of the range of variation existing within populations and its adaptive value under changing ecological conditions. Being quantitative, such studies have enabled us to calculate the selective advantage or disadvantage of particular variants and hence their potential for evolution. Polymorphic variation in species is particularly useful in ecological studies, since the different forms are often distinct and easily recognised in the field. Thus the white-lipped snail, *Cepaea hortensis*, has two forms of shell, plain yellow with no bands (unbanded), and yellow with a number of dark brown bands (usually five) running round it from spire to lip (Fig. 9.5 p. 170). In an area of Portland (Dorset) where the snails were common, it was found from a comparison of living snails and the shells of those predated by thrushes that there was a consistent selection of about eight per cent against banded shells, irrespective of whether they were from high or low-banded populations [1.4]. The reasons for this result are not clear, but it may well be to do with the searching image of the birds, shell banding being associated with food and therefore selected preferentially irrespective of rarity.

Of all selective agents known, the most powerful is undoubtedly man. As a result of his activities we now have DDT-resistant cockroaches and mosquitoes, warfarin-resistant rats, and greenfly resistant to the insecticide malathion in our glasshouses.

As was pointed out earlier, the study of ecology has strong links with that of evolution, and nowhere is this clearer than in the field of intraspecific competition. The wider implications of such studies, their relationship with the formation of new species, and their technological implications will be considered further in later chapters.

Relationships between species

By contrast with intraspecific interactions, the relationships between species (**interspecific**) are more complex and their study leads us to consider some of the most fundamental features of plant and animal communities. No ecosystem can exist for long unless it is founded on a balanced set of food relationships. That is to say, over a period of time, the level of production must equal or exceed that of consumption. All energy is ultimately derived from the sun, and through their autotrophic method of feeding (photosynthesis) green plants are able to capture between one and five per cent of the energy that falls upon them in the form of light. They thus occupy the position of **primary producers**. Herbivores browse on plants (**primary consumers**), carnivores eat herbivores (**secondary consumers**), while small carnivores may be preyed upon by larger carnivores (**tertiary consumers**). In this way, energy and other requirements for life are passed along food chains from one stage to the next, each stage being known as a **trophic level**. But the food relationships of communities are by no means a static system (as they are so often depicted). Most animals eat a wide variety of foods according to their availability, and since population densities may fluctuate considerably from year to year, the pressure exerted upon particular food supplies varies too. Moreover, it must be remembered that young animals frequently require different foods from adults, so the age of a population may affect the range of its food requirements. The quantitative aspects of energy flow through ecosystems and a more intimate examination of food relationships can be left until later (Chapter 4). Suffice it to add here that in both populations and communities there is intense competition for food, but whereas in populations this is centred on the limited spectrum of requirements of a single species, in communities the range of foodstuffs required for survival is far greater and the relationships between the different contributors more complex.

Competition for living space

In populations, as we have seen, there is strong competition for somewhere to live. This leads to differential survival of some variants and the evolution of new forms. In communities too, similar competition between species results in a dynamic situation promoting change. The colonisation of a previously uninhabited locality depends partly on the chance arrival of propagules and also on their ability to establish themselves quickly in the face of fluctuating environmental conditions and competition from other competitors. Primary plant colonists of bare soil are usually annuals, and if allowed to survive many of these will later be replaced by perennials (both herbaceous and woody). This process is known as **succession**. But it is not just a question of one species ousting another. In the process profound changes may take place in the environment, such as an increase in the humus and water content of the soil and a modification of the microclimate above it. When interspecific competition brings about alterations in the structure of a plant community, it also influences the animal species that depend upon it. The principle is well illustrated by observations of a piece of burnt woodland which extended over a period of five years [1.5]. Following the burning, the first plant colonists were ragwort, *Senecio jacobaea*, and rosebay willowherb, *Epilobium angustifolium* (Fig. 1.3), which shared the habitat more or less equally between them. It was noticeable that many of the ragwort plants were colonised by the familiar black and yellow banded larvae of the cinnabar moth, *Callimorpha jacobaeae* (Fig. 1.5). There was also a significant increase in larvae of the small elephant hawk-moth, *Deilephila porcellus*, feeding on the *Epilobium*. Subsequent years saw the rosebay greatly extending its range at the expense of the ragwort, no doubt due to its capacity for rapid vegetative growth by means of underground rhizomes. The decline of the ragwort heralded a reduction in the numbers of cinnabar moth larvae but an increase in the density of small elephant hawk moths. The eventual outcome of such changes would have been some sort of equilibrium between the requirements of the colonists themselves and their environment. This is known as a **climax**. In the example just quoted, it is likely that, if left undisturbed, the end product would have been a wood. A classic instance of such a succession is provided by part of the Broadback wheat field at Rothamsted Experimental Station, Harpenden, which has been left unharvested since 1882.

Fig. 1.5 Larvae of the cinnabar moth, *Callimorpha jacobaeae*, feeding on ragwort, *Senecio jacobaea* (photograph Frank W. Lane)

Records show that it soon became a dense thicket and it is now a fully grown oak wood. Associated with this change have been corresponding alterations in the animal population; for instance, the locality now supports a typical woodland community of insects.

Adaptation and adaptability

Throughout this chapter, stress has been laid on one of the fundamental properties of populations and communities, namely that they represent dynamic systems in a constant state of change. Whether the foothold achieved by a colonising species is ephemeral or more long-lasting is to a large extent determined by its **adaptability**. Adaptations in plants and animals manifest themselves in many ways. We tend to think of them in terms of structural features such as the transport systems (xylem and phloem) of plants or the feeding mechanisms of animals. However, ultimately they are all physiological, being concerned with the effective continuance of vital processes such as photosynthesis and reproduction. We have seen how the capacity of plant species to exist at different average levels of illumination can play a profound part in influencing spatial distribution. Similarly, the foods of animals are restricted to a limited range of possible alternatives and this is reflected in the structure of their feeding mechanisms which therefore provide a useful guide to their diet and likely habitats.

The place in which an organisms lives is its **habitat**. Because of its limited adaptability, a species tends to inhabit some places but not others, so all species have habitat preferences. The **ecological niche** occupied by a species indicates the economic position of that species within the community and is therefore a useful term for defining food relationships. Among plants, we may distinguish between saprophytes, such as many fungi, and autotrophs relying on photosynthesis. Among the autotrophs we have sun and shade species. Among animal communities we find herbivores, carnivores, and insectivores, each group occupying a distinct position within a particular ecosystem. As Elton has charmingly put it, 'When an ecologist says "There goes a badger", he should include in his thoughts some definite idea of the animal's place in the community to which it belongs, just as if he had said "there goes the vicar"' [1.1].

Some organisms such as nematode worms show remarkable powers of adaptation and seem to be able to exist almost anywhere, their habitats ranging from soil to the human intestine and the beer mats in pubs! Others are much more limited in their requirements, like the flatworms (Cestoda) which inhabit the alimentary canals of many animals but are unable to exist outside them. A species which remains as uncommitted as possible, so that it can adapt to a wide range of environmental conditions, is more likely to survive. However, once a species is committed to some specialised sort of existence, such as an extreme form of parasitism, it is unlikely to be able to readapt and if conditions change beyond its limited powers of adjustment, extinction will result.

Earlier in this chapter the point was made that one of the main difficulties in studying ecology derives from the fact that communities of plants and animals are so complex. Patterns are not easy to discern and hence in analysing them it is often difficult to decide which questions to ask. The Bureau of Animal Population at Oxford University carried out an ecological survey, based mainly on Wytham Woods near Oxford, which extended over a period of twenty years and involved a large number of researchers. Elton [1.6] concludes that, 'To know thoroughly these few square miles would really need more days and nights and less sleep than any one person can contemplate'.

One of the advantages of adopting a holist approach is that it reveals living populations and communities as dynamic systems subject to constant change. Moreover, it helps us to realise how the interaction of such fundamental ecological processes as reproduction, dispersal, colonisation, and competition serve to bring about a balance within and between different ecosystems. The result is what we see around us, namely the *spatial distribution* of organisms, and this therefore provides a logical starting point for any study of ecology.

Summary

1 Populations and communities of living organisms interact with one another and their physical environment in a complex manner which is usually difficult to interpret completely.

2   The survival of organisms within a habitat depends upon their ability to adapt to the particular conditions that exist there. All adaptations are ultimately physiological.

3   Within a habitat, the interactions between living things represent systems (ecosystems) consisting of distinct but related entities which are linked together and mutually dependent.

4   Within ecosystems, the basic unit is the species. Defining a species is not easy. A number of criteria can be applied not all of which are applicable in a particular situation.

5   Problems of identification are increased by the occurrence of variation due to the effects of the environment which can give rise to gradients (clines). Genetic influences such as polygenic inheritance and polymorphism are also important causes of variation.

6   The existence of variation has resulted in the concept of the polytypic species in which a certain degree of variety is accepted. The application of this principle differs somewhat in plants and animals.

7   The effective colonisation of a habitat depends on a number of factors such as reproduction and dispersal, competition and adaptability resulting from variation.

8   Relationships between species provide the basis of food chains and a balance of producers and consumers. Such systems are in a constant state of fluctuation.

9   Competition for living space results in the process of succession, the theoretical end point being a climax. The rate at which succession occurs varies greatly in different communities.

10 The terms habitat and niche are both useful in the study of ecology. A habitat is the *place* where an organism lives, a niche represents its *economic position* within a community.

References

1.1   Elton, C. S. (1927) *Animal Ecology*, Sidgwick and Jackson
1.2   Crowson, R. A. (1970) *Classification and Biology*, Heinemann
1.3   Mayr, E. (1964) *Systematics and the Origin of Species*, New York, Dover Publications
1.4   Dowdeswell, W. H. (1975) *The Mechanism of Evolution*, 4th edn, Heinemann
1.5   Dowdeswell, W. H. (1980). *Report of the Bath Natural History Society*, 994–997
1.6   Elton, C. S. (1966) *The Pattern of Animal Communities*, Methuen

2

Patterns and distribution

In relatively elementary ecological studies the starting point will almost certainly be a community of organisms that is readily accessible, such as in an area of grassland, a hedgerow, or the bed of a stream. A convenient entry to such studies is to adopt a holist approach, viewing the habitat in question as a single ecological entity. We can then ask certain fundamental questions about it, such as, 'What are the general features of plant and animal distribution?' and 'What are the principal factors which are most likely to have brought the existing situation about?'. Most distributional patterns are complex, and in attempting to understand them we have to adopt a reductionist approach, selecting small portions and studying them in some detail. Using the techniques of synthesis and analysis discussed in the previous chapter, we can then hope to build up a composite picture to explain at least some of the features of the pattern as a whole.

Habitat preferences

As we have seen, under natural conditions plants and animals are never distributed at random, the different species exhibiting characteristic **habitat preferences**. Some plants are shade-loving and shun full sunlight; many nymphs of mayflies show a preference for the undersurface of stones in the midst of a current but some prefer to burrow in the mud where the water is static. The principle can be illustrated by a simple laboratory experiment using the flour beetle, *Tribolium confusum*. A small chamber such as a 500 cm³ beaker or a plastic box is set up, containing equal volumes of three different media: flour, crushed oats, and sawdust. The media can be kept separate using pieces of cardboard (Fig. 2.1). The beetles are released at the centre, the cardboard divisions having first been removed so as to permit uninterrupted movement. After a few days during which the animals have been allowed to distribute themselves, the cardboard divisions are re-inserted and the populations in the three media are sampled using a large spoon. We now have data for the spatial distribution of *Tribolium* and we

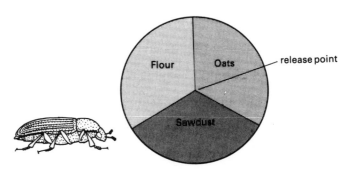

Fig. 2.1 Experiment to study habitat preference in the flour beetle, *Tribolium confusum*

can proceed to test the null hypothesis that the individuals have distributed themselves equally, using a chi-squared (χ^2) test. The procedure is described in Chapter 7 (p. 138).

As might be expected, the beetles are found to show strong habitat preferences in the order, flour, oats, sawdust. However, if the experiment is left for several weeks it will be seen that the distribution pattern fluctuates somewhat, the original clear-cut preferences being partially obscured. Clearly, secondary influences are involved, among them competition within the three beetle populations. Incidentally, this exercise also provides a good opportunity for gaining experience of the influence of sampling error on the significance of the results achieved due to the size of the samples and the number taken.

Physical factors and plant distribution

We have seen in Chapter 1 (p. 7) how light is one of the most powerful abiotic factors influencing the distribution of plants such as the rosbay willowherb and ragwort. Such an effect is commonplace and can readily be observed among species growing at the margins of woods or near individual trees or bushes which cast appreciable shade. Humidity is another factor exerting a profound effect on the spatial distribution of plants. If we study the colonists of any stream or pond it will be obvious as we approach the interface between water and land that the pattern of the plant species is changing. The nature of the change and its abruptness will vary with circumstances such as the fall of the ground, nature of the soil, depth of the water and so forth. In general, it will follow the sequence:

Plant submerged,	Some parts aerial,	All parts aerial,
rooted in water	rooted in water	rooted on land

Water ⟶ Land

Such a transition from water to land is known as a **hydrosere**. A similar situation can occur if the change is from dry to wetter conditions, in which case we have a **xerosere**. This can often be observed on old roofs, particularly if the tiles are of stone (Fig. 2.2) and overhung by trees. The falling leaves and the humus resulting from their decay often accumulate unevenly, resulting in patches of moisture amid a dry surround. The resulting xerosere is then:

Dry, little humus	Moister, some humus	Moister, much humus
Lichens ⟶	Mosses ⟶	Ferns
e.g. *Xanthoria* spp	e.g. Hairy screw moss	e.g. Common polypody
	Tortula ruralis	*Polypodium vulgare*

Dry ⟶ Moist

The study of old roofs and walls serves to illustrate the point made earlier, that while the nature of the physical environment determines the pattern of colonisation by plants and animals, the process of colonisation can itself exert radical changes on the nature of the environment. Thus on a roof, any dead remains of

Fig. 2.2 The common fern, *Polypodium vulgare*, colonising an old stone roof

lichens or mosses will provide significant local accumulations of humus, and these in turn will aid the development of the xerosere.

Other powerful factors determining the distribution of plants are those concerned with the nature of the soil (**edaphic factors**), such as mineral content and pH. Of increasing interest during recent years have been those elements such as the heavy metals which can act as pollutants and we shall be considering them further later in the chapter (p. 24).

Physical factors and animal distribution

The influence of physical factors on the distribution of animal populations is more difficult to discern, as these factors seldom exert their effects alone. Moreover, animals are more variable than plants in their reactions to environmental change. One of the nearest approaches to a study of the influence of a single abiotic factor on an animal, is that on the tolerance of the water louse, *Asellus*, to different levels of calcium in the tarns of the Lake District. Two different sets of data (from T. B. Reynoldson and H. P. Moon) have been summarised by Macan [2.1] and are shown in Fig. 2.3. In spite of considerable variation, there is clearly some common ground between the two groups of observations. Thus, where the calcium concentration was less than 5 mg l^{-1}, *Asellus* rarely occurred, but between the levels of 7 and 12.5 mg l^{-1} it might or might not be present. Where the concentration was more than 12.5 mg l^{-1}, *Asellus* appeared to be present universally, a finding confirmed by studies in other parts of the world where the

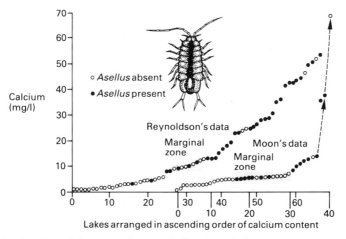

Fig. 2.3 Distribution of the water louse, *Asellus*, in relation to the calcium content of the water

animal occurs. As Macan concludes, such data need to be interpreted with considerable caution since we may well be dealing with an organism which is capable of wide variations in calcium tolerance. Moreover, in seeking an interpretation of data such as those in Fig. 2.3, we need to take account of the animal's behaviour and its considerable capacity to move over land. Thus it may well be that individuals faced with an inimical calcium environment in one place stand a good chance of being able to move elsewhere in search of conditions that are more favourable.

The simultaneous influence of more than one physical variable is typified by the situations encountered when studying the ecology of grass tufts and the animals such as the larvae of the meadow brown butterfly, *Maniola jurtina* [2.2], that inhabit them. These larvae are highly palatable to diurnal predators such as birds, which no doubt accounts for the fact that they are most active at night. In order to sample such a population after dark it was desirable to be able to predict the circumstances in which a population was likely to be at its most active. Under adverse conditions, the animals remained in the depths of the grass and defied any attempts to capture them by means of a sweep net. For successful collecting, a minimum temperature of 10°C and a high humidity resulting from a heavy dew or light rain were necessary. If either of these requirements were not met, the larvae invariably remained inactive.

Physical factors and animal activity

The activities of most animals living in or on soil are controlled by the combined effects of humidity, temperature, and light intensity. Slugs, for example, have well developed eyes and a rate of metabolism which responds readily to the amount of heat obtainable from the atmosphere. Humidity is also important, since their skin is covered with mucus and they lose a good deal of water by evaporation and through the hydrated slime that lubricates their path when crawling. In a study of

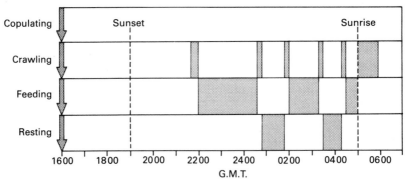

Fig. 2.4 Activities of a slug (*Agriolimax reticulatus*) between 21.45 and 05.45 GMT. Shaded areas denote the duration of different kinds of activity (after Newell)

the behaviour pattern of the slug, *Agriolimax reticulatus*, Newell [2.3] showed that surface crawling was normally limited to the hours of darkness. Hence it seemed reasonable to assume that the environmental stimuli to which the species responds might be correlated with the changes associated with nightfall and daybreak. Time-lapse photographic records were obtained of the behaviour of individual slugs at different times of the day and night, measurements being made of the soil temperature (using a thermistor probe, see p. 104) and light (using a photoelectric cell and automatic recorder, see p. 106). The record for a typical slug is summarised in Fig. 2.4. From such observations, Newell concluded that activity in the soil must be initiated by some kind of internal rhythm. When the slugs burrow out of the soil and reach the surface they are deterred if the eyes detect bright light and do not emerge until the level of illumination declines to a low threshold value, usually after sunset. As might be expected, the precise time when this threshold is reached varies with the weather and the aspect of the habitat. In order to obtain reasonably consistent and comparable results it was found necessary to use a relatively plant-free area with no obstructions in the immediate vicinity. By dawn, many slugs will already have returned to the soil but those that have not will be stimulated by the increasing light to rapid movement homewards. Thus, just as the animals occupying grass tufts (lepidopterous larvae, snails, ants, and others) move up and down the grass stems in response to changing environmental conditions, so soil inhabiting species such as slugs perform similar movements within their terrain. Both exhibit a relatively clear-cut pattern of day and night distributions, alternating continually between them.

Dynamic responses to changes in temperature and illumination are even more marked among planktonic organisms inhabiting all kinds of static water. In ponds and lakes the various kinds of water fleas (Cladocera) such as *Daphnia* and *Simocephalus* often abound in the brightly illuminated zone. These populations can easily be sampled with a plankton net and the individual animals isolated by filtering through a Büchner funnel. During warm weather continuous sampling reveals a marked variation in spatial distribution at different times of the day and night. Changes take place according to a distinct pattern, the majority of the population coming to the surface at night and moving downwards during daytime as the light intensity increases.

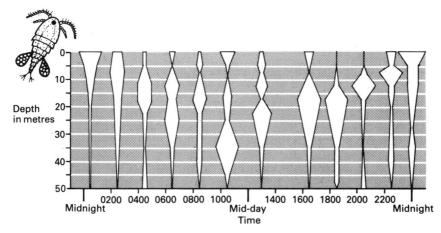

Fig. 2.5 Daily migration of the copepod, *Cyclops strenuus*, in Lake Windermere. The numbers at each depth are a percentage of the total sample (after Ullyott)

A typical example of this cyclical change in distribution is that of the copepod, *Cyclops strenuus*, in Lake Windermere (Fig. 2.5). At about midday the surface waters are almost depleted, the population reaching its maximum abundance at a depth of 35 metres. But as day draws to a close and the illumination gradually declines, an upward migration occurs, attaining a maximum at midnight. As light returns, a downward movement begins once more. Clearly, the movement of the plankton and the position of the sun are closely related, but whether light is the only factor influencing the spatial distribution of planktonic organisms such as *Cyclops* is still an open question. The situation evidently becomes more complicated in winter when the temperature of the water drops to 5°C or below. Studies of *Daphnia pulex* suggest that the lower the temperature, the less is the effect of light in influencing distribution.

Erosion by wind and water

Another powerful environmental influence controlling the nature and distribution of the plants and animals colonising a habitat is the eroding force of wind and water. Examination of any exposed area of moorland will illustrate the fact that most species of trees are unable to withstand the searing effects of prolonged wind and rain. Even such hardy species as the rowan or mountain ash, *Sorbus aucuparia*, are usually to be found colonising gulleys or combes which afford some protection from the prevailing wind.

The effects of flow velocity on animal distribution are equally striking but more localised, as in the rapids of streams and rivers. Edington studied the larvae of two species of caddis fly inhabiting streams in northern England [2.4]. The larva of *Hydropsyche siltalai* is mostly a detritus feeder and spins a rigid net which filters out small particles just like a plankton net. The larva of *Plectrocnemia conspersa* is carnivorous and uses its rather flimsy net to capture invertebrate prey such as small mayfly nymphs, either while the prey is walking about or by its being washed into the net by the currents. Both species of caddis fly larva concentrate their nets on the tops of stones or boulders and this enables a miniature flow-meter to be used to compare the velocity of the water at their mouths. Nets of *Hydropsyche* were found in current speeds of between 15 and 100 cm s⁻¹, but were commonest in the fastest water. *Plectrocnemia* preferred lower water velocities of between 0 and 20 cm s⁻¹, 95 per cent being found in speeds less than 10 cm s⁻¹. Edington found that if the flimsy nets of *Plectrocnemia* were transferred to rapids upstream they were quickly ruptured and soon became non-functional. This was confirmed in the laboratory, where a water velocity of 25 cm s⁻¹ was sufficient to destroy a net. The rigid net of *Hydropsyche* however, proved to be well adapted for currents of all speeds. Experimental evidence of this was provided by laboratory investigations relating variations in flow velocity to success in net-building. The results are summarised in Table 2.1.

Table 2.1 Relationship between current speed and net-building success (%) in the larvae of two species of caddis fly. Figures in brackets denote size of samples tested. (After Edington)

	Current speed (cm s⁻¹)		
Caddis fly species	10	15	20
Hydropsyche siltalai	20 (25)	48 (27)	73 (22)
Plectrocnemia conspersa	72 (25)	50 (22)	4 (26)

The laboratory findings thus accorded closely with the distributional pattern occurring under natural conditions, the larvae of *Hydropsyche* showing a preference for rapids while those of *Plectrocnemia* predominated in pools. Adaptation involved not only the form of the net but also larval behaviour patterns associated with its siting and construction. As Edington points out, it is thus possible to speak of low-velocity and high-velocity species. This would be a profitable field for the investigation of other groups of aquatic invertebrates such as mayfly nymphs which exhibit marked differences in habitat preference, some of which are related to the current.

Some of the most extreme examples of the effects of wind and water on the distribution of plant and animal populations are to be found on the sea shore, where the daily rise and fall of the tide pose problems concerned with desiccation, the maintenance of gaseous exchange, and protection from predators. Variation in tolerance of exposure in different species results in their characteristic **zonation**, described in Chapter 11.

Dispersion and association of species

From the examples discussed earlier, certain generalisations can be made about the spatial distribution of plant and animal populations. The way in which they are distributed usually conforms approximately to a few definable patterns. For instance, distribution may be **random**: the probability of an individual occurring in one place is the same as of its occurring in another. Evidence of the powerful selective influence of environmental variables and competition between species suggests that such a pattern is likely to be restricted to instances such as the wind dispersal of plant fruits and seeds and the seedlings that first appear from them. However, any degree of randomness that may occur is unlikely to remain for long in the face of strong differential survival. Dispersion may be **uniform**. In such a distribution, a species does not occur universally in an area, but when present it is distributed uniformly. Such situations seldom occur in wild populations, but are typical of monocrop cultivation by man, such as in fields of corn. The type of distribution occurring most commonly in natural populations involves groups of individuals scattered somewhat irregularly throughout a habitat. Such a pattern is said to be **clumped**.

The clumping of individuals of a species within a particular habitat frequently results from particular environmental conditions. For instance, the dry, acid soils of heather moors favour colonisation by such plants as ling, *Calluna vulgaris*, and bell heather, *Erica cinerea* (Fig. 12.2 p. 237), which tend to predominate over all other species. In order to define such a situation more precisely, botanists sometimes add the termination '*etum*'. Hence a moor dominated by *Calluna* is referred to as a *Callunetum*. This implies not only that ling is present but that other species associated with such conditions are likely to be there too. An alternative means of defining a community is in terms of **indicator species** – those which are characteristic colonists of a particular kind of habitat. Thus *Calluna vulgaris* would qualify as an indicator species for the ecological conditions that comprise a heather moor.

Evidence of clumping and the existence of indicator species is also widespread among animal populations. On a rocky shore, the various kinds of seaweeds (e.g. *Fucus*) are zoned according to their tolerance of exposure, as are gastropod molluscs associated with them, such as the different species of *Littorina* (see Chapter 11 p. 210).

Species diversity and distribution

A general characteristic of all natural communities is that they tend to contain comparatively few species that are common and a much larger number that are relatively uncommon. This principle is well illustrated in dry moorland, which is usually dominated by two species, ling and bell heather. However, the total number of plant species represented may well amount to several hundred. The diversity of any community is a function of the number of different species represented and the numbers of individuals belonging to each species. The more mature it becomes the greater its diversity. However, from this it does not necessarily follow that the fewer the number of species, the earlier the stage of colonisation.

In general, the life sequence of a community passes through five phases:

(i) The initial stage is when the primary colonists are arriving. These are usually propagules such as plant seeds and spores, also fragments of plants that are capable of vegetative propagation which are blown there by the wind or transported by animals.

(ii) There follows a period of instability marked by intense competition between species, and reactions, favourable or otherwise, to the physical environment. During this phase the number of plant species surviving is often reduced and the first animal colonists arrive, mostly insects.

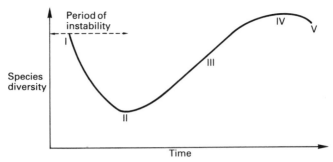

Fig. 2.6 Schematic representation of five phases in the species diversity of a community: I – colonisation; II – instability; III – increase; IV – climax; V – ageing (see text)

(iii) The community now enters a period of gradually increasing maturity and the number of plant and animal species increases.
(iv) Eventually, a stage of maturity is reached which is usually referred to as a **climax** (see also p. 67). At this point, species diversity reaches a maximum.
(v) The process of ageing brings about a gradual reduction in the community through death and decay, and the number of species begins to decline.

The sequence of events outlined above is summarised schematically in Fig. 2.6.

Various methods have been devised for quantifying the diversity of a plant community (**diversity index**). All of them are based on the relationship between the total number of plants present and the number of individuals per species, but the different approaches vary in the extent to which they are biased in favour of the relatively few common species or the more numerous rarer ones. Detailed discussion of the different formulae that have been proposed is outside the scope of this book but is readily available elsewhere [2.6, 2.7, 2.8]. One of the simplest to calculate is the Simpson Index which is given by the formula:

$$D = \frac{N(N-1)}{\sum n(n-1)}$$

Where D = diversity index
N = total number of individual plants
n = number of individuals per species
$\sum$ = sum

The procedure involves noting every plant in the area under investigation. If two communities are to be compared, say one in the initial colonisation phase and the other which is on the increase, the same area of ground, for example a 5 m square, must be studied for each. The precise naming of the different species is not necessary, provided a means of identification is used such as a letter, number or pseudonym. A typical example will illustrate the procedure.

An area of soil which had been recently cleared of vegetation was studied after one year's colonisation, and the following flowering plants were found in an area five metres square:

Species	Number
Daisy (*Bellis perennis*)	15
Scentless mayweed (*Matricaria perforata*)	10
Groundsell (*Senecio vulgaris*)	104
Ragwort (*Senecio jacobaea*)	42
Hairy bittercress (*Cardamine hirsuta*)	55
Fat hen (*Chenopodium album*)	22
Nettle (*Urtica dioica*)	9
Forget-me-not (*Myosotis arvensis*)	7
Field speedwell (*Veronica persica*)	17
	281

$$D = \frac{281(281 - 1)}{15(15 - 1) + 10(10 - 1) \text{ etc.}} = \frac{78\,680}{16\,552}$$

$$= 4.8$$

On visiting the area a year later, the diversity index was found to have dropped to 3.9, suggesting that the community was in the phase of instability.

Wratten and Fry [2.5] quote a simple and convenient method for determining animal diversity in freshwater habitats such as ponds, streams, and rivers. Either a number of samples can be collected from different localities or a few taken from the same area, if a study is required in more detail. In the laboratory the organisms in each sample are transferred to white plastic photographic dishes with parallel lines drawn on the bottom roughly two centimetres apart. Before transferring the animals by a wide-mouthed pipette or some other means, it is important to mix the sample gently to ensure randomisation. The animals are then arranged in rows with a paint brush and classified into runs of the same species. Thus, in a stream, we might find the nymphs of the mayfly, *Ephemerella ignita*, the snail, *Hydrobia jenkinsi*, and the planarian, *Polycelis cornuta*. For convenience, these can be labelled A, B, and C respectively. In a particular line of the three species their distribution could be as follows:

$$\underset{1}{\underbrace{A\ A}}\ \underset{2}{\underbrace{B\ B}}\ \underset{3}{\underbrace{A\ A}}\ \underset{4}{\underbrace{C\ C}}\ \underset{5}{B}\ \underset{6}{\underbrace{A\ A}}\ \underset{7}{C}\ \underset{8}{\underbrace{B\ B}}\ \underset{9}{\underbrace{C\ C}}$$

There are thus 9 runs and 16 animals.

A simple way of scoring diversity is to record 1 for the first individual and thereafter 1 for each change. Applying this procedure we then have:

A	A	B	B	A	A	C	C	B	A	A	C	B	B	C	C = 16
1		1		1		1		1	1		1	1		1	= 9

The Sequential Comparison Index (index of diversity) is then given by:

$$\text{SCI} = \frac{\text{number of runs}}{\text{number of organisms}}$$

The size of the samples examined is an important consideration and for a reliable estimate of diversity at least 200 animals should have been recorded. Where mobile species are involved, it may be necessary to apply a mild anaesthetic, such as carbon dioxide, to the water, or to cool the samples in a refrigerator before examining them. From such studies a number of interesting questions can arise. For instance, to what extent does the total number of species present in a particular habitat correlate with the estimate of diversity? And to what extent do high and low diversity scores throw light on the habitat preferences of the organisms concerned?

Pollution and spatial distribution

Of recent years we have become increasingly aware of the influence of human activities on the ecology of natural plant and animal populations. Some of the more widespread effects are now well documented, for example, the effect of the sulphur dioxide emitted into the atmosphere in industrial areas as a result of the burning of fossil fuels. Evidently, lichens are extremely sensitive to the gas, and in regions of high sulphur dioxide concentration the colonisation of tree trunks by lichens is reduced or absent. This has had profound effects on many species of moth, such as the peppered moth, *Biston betularia*, which formerly relied on a lichen-covered background to provide concealment from bird predators when at rest. The phenomenon of industrial melanism, in which rare, dark (melanic) varieties have found themselves at a selective advantage and increased in number since they match the lichen-free background more effectively than the typical forms, has been recorded in more than one hundred different species in Britain alone, besides being widespread elsewhere.

Bradshaw [2.7] and his co-workers have investigated the peculiar distribution of plant species in the vicinity of old Welsh mine workings, where the ores of heavy metals such as zinc, copper, and lead were once mined. A characteristic feature of the landscape in these areas is the old spoil tips, which are largely devoid of vegetation on account of the poisonous nature of minerals in the soil. Yet, here and there, a few clumps of grasses such as *Agrostis tenuis*, *Festuca ovina*, and *Anthoxanthum odoratum* (Fig. 2.7) survive. Bradshaw showed that such species

Fig. 2.7   Discontinuous distribution of heavy-metal tolerant species of grass in the vicinity of a lead and zinc mine (Nant, Wales) (photograph Professor A. D. Bradshaw)

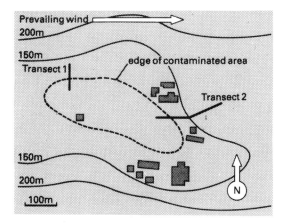

Fig. 2.8 Plan of the mine area studied by McNeilly

had developed tolerance to the metal pollutants to which they were exposed and that this capacity was inherited. We now know that heavy-metal tolerance is not confined to grasses but occurs in such diverse species as the bladder campion, *Silene vulgaris*, and the ribwort plantain, *Plantago lanceolata*. However, the development of tolerance is by no means a characteristic of all plants growing near mines, and species such as the grass, *Dactylis glomerata*, and the bird's foot trefoil, *Lotus corniculatus*, have singularly failed to adapt to the polluted conditions.

McNeilly [2.8] studied the distribution of metal tolerance among plants of the grass, *Agrostis tenuis*, in the vicinity of a small copper mine in North Wales. His tests showed that the soil close to the mine was so heavily polluted that plants taken from uncontaminated areas elsewhere were unable to survive in it. Even at sub-lethal concentrations of copper, root growth was found to be severely retarded. The degree of tolerance exhibited by *Agrostis* was determined in the laboratory by estimating root growth in different concentrations of copper in a standard time, the results being expressed as an index of tolerance. A plan of the area investigated is shown in Fig. 2.8.

McNeilly examined two transects (see Fig. 2.8), one at the west end of the contaminated area and across the prevailing wind, the other further east and along the line of the wind. He estimated the copper tolerance both of adult plants and of the populations derived from their seeds. The results are summarised in Fig. 2.9. In the first transect there was an abrupt change in the tolerance of *Agrostis* at the boundary of the mine between toxic and non-toxic soils. In the second transect downwind, tolerance was discontinuous and gradually faded out over a distance of 150 metres of relatively uncontaminated ground. Three particular points of interest arise from these studies. The fact that the transects across the prevailing wind and downwind produced such different results suggests that wind-blown pollen carrying the genes that control tolerance must have played an important part in determining spatial distribution. Again, the abrupt change in tolerance (in only a metre) between non-mine and mine populations in the first transect suggests that powerful selective forces have influenced the survival of plants that are adapted to the particular conditions, thus ensuring that their beneficial characteristics are transmitted via their seed to the next generation. Finally, the

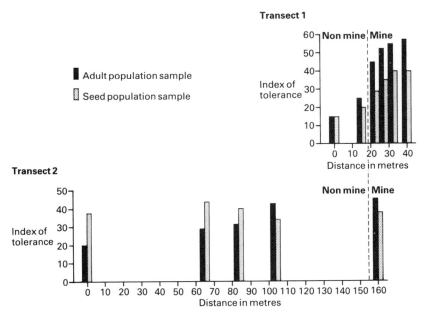

Fig. 2.9 Copper tolerance of populations of the grass, *Agrostis tenuis* in the vicinity of a Welsh mine (after McNeilly)

difference in mean tolerance between adult and seed populations again indicates the action of natural selection during the life-time of the young plants.

The fact that such powerful selective forces are acting at the boundary of the polluted area raises the possibility that changes in the tolerance of the plants may have been achieved quite rapidly, perhaps during the period of existence of the mine. But, as Bradshaw has pointed out, estimates based on this supposition could be misleading, for it seems likely that some contamination of the area may have existed before the mining began and with it a corresponding tolerance in the local plants. Some evidence of the speed with which tolerance can be evolved has been derived from grasses such as *Agrostis canina* and *Festuca ovina* growing under newly erected galvanised wire fences. Evidently, an appreciable degree of zinc tolerance can be achieved in less than thirty years [2.7]. The discovery of these pollution-tolerant strains of grasses has also had great practical consequences, for it has enabled derelict regions of contaminated ground left over from former mine workings and slag heaps to be replanted with vegetation, converting them into areas with considerable aesthetic and amenity value.

In conclusion, it is worth pointing out that while the central theme of this chapter has been the spatial distributions of species and the factors affecting them, we began with a strictly ecological approach and finished with a genetic and evolutionary one. That such a change of emphasis in a single topic can occur so readily, serves to underline the close relationship that exists between the two areas of study (see Fig. 1.1 p. 2). Indeed, in the interests of fostering the union between them, it has proved desirable to coin the term **ecological genetics** [2.9], a move which has opened up a fresh approach to the study of populations and the processes which underlie their patterns of distribution.

Summary

1 Plants and animals show distinct habitat preferences which can be studied experimentally and in the field.

2 Physical factors such as light and temperature play a powerful part in influencing plant and animal distribution. They seldom act alone but do so in conjunction with other factors. They also frequently determine the pattern and extent of animal activity.

3 Erosion by wind and water may also have profound effects on spatial distribution.

4 The dispersion of organisms in the wild state frequently conforms to a few definable patterns. Estimates of species diversity (diversity index) can be a useful aid in the analysis of such patterns.

5 Pollution can exert important effects on the distribution of both plants and animals. Experimental evidence indicates the extent of these effects and how they may have come about.

References

2.1 Macan, T. T. (1963) *Freshwater Ecology*, Longmans

2.2 Dowdeswell, W. H. (1981) *The Life of the Meadow Brown*, Heinemann

2.3 Newell, P. F. (1968) 'The measurement of light and temperature as factors controlling the surface activity of the slug, *Agriolimax reticulatus*, (Müller)' in *The Measurement of Environmental Factors in Terrestrial Ecology*, Blackwell

2.4 Edington, J. M. (1968) 'Habitat preferences in net-spinning caddis larvae with special reference to the influence of water velocity', *Journal of Animal Ecology*, **37**, 675–92

2.5 Wratten, S. F. and Fry, G. L. A. (1980) *Field and Laboratory Exercises in Ecology*, Arnold

2.6 Odum, E. P. (1975) *Ecology*, 2nd edn, Holt, Rinehart and Winston

2.7 Bradshaw, A. D., McNeilly, T. S., and Gregory, R. P. G. (1965) 'Industrialisation, evolution and the development of heavy metal tolerance in plants', *5th Symposium of the British Ecological Society*, 327–43, Blackwell

2.8 McNeilly, T. S. (1968) 'Evolution in closely adjacent plant populations', *Heredity*, **23**, 99–108

2.9 Ford, E. B. (1975) *Ecological Genetics*, 4th edn, Chapman and Hall

3

The dynamics of populations

The fecundity of living organisms varies greatly from one species to another. While mammals and birds, with their high level of parental care, produce only a few fertilised eggs each year, fishes such as the ling, *Molva vulgaris*, have been estimated to lay around 160 million eggs annually. Again, weeds like the rosebay willow herb, *Epilobium angustifolium*, produce about 80 000 seeds per flower, while a single mushroom can generate as many as three million spores.

Population size

The size of a population, excluding the effects of immigration and emigration, depends upon a balance between reproductive survival and death rate. This is shown diagrammatically in Fig. 3.1. Of the vast number of propagules produced by most plant and animal species, few (far less than one per cent) survive to the stage of reproductive maturity. Some will succumb to an unfavourable physical environment, others die of disease, but the majority will provide the food for a host of predators. These variables, and others related to them, are by no means constant in their effects, and their relative influence on survival varies considerably from one season to the next.

The numbers in a population of a particular species are usually expressed as its **density**. That is to say, the number of individuals per unit area of habitat. For diverse habitats such as streams, density can be a rather crude parameter, for it will include some zones where the species attains its maximum numbers and others where it is almost absent. To have any precise meaning, figures for the density of a population need to be closely related to its spatial distribution. A variety of methods are now available for making estimates of population numbers and these are discussed in Chapter 5.

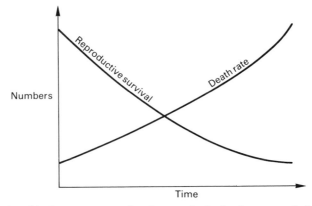

Fig. 3.1 Relationship between reproductive survival, death rate, and size of a population

The growth of populations

If a population of organisms is allowed to grow in isolation from any other population which might compete with it, the pattern of increase is likely to conform to an S-shaped (sigmoid) curve as shown in Fig. 3.2. The sequence of events can be divided into three distinct phases. At first growth will be rapid, multiplication being exponential (geometric). Later, the rate of increase will slow down as a depletion of resources and other limiting factors begin to exert their effects. Finally, a steady state will be achieved in which the density remains almost constant and no further increase in numbers occurs.

Clatworthy and Harper [3.1] investigated experimentally the pattern of population increase in the common duckweed, *Lemna minor*, when grown on its own and in competition with another species. *Lemna* is a small floating plant which is common in ponds and which multiplies by the division of its leaves. When a few of these plants were added to pond water in a glass beaker there was an initial period of exponential growth, based largely on the food resources available within the plants themselves. However, as the fronds multiplied and spaced out over the surface of the culture, a mat began to form. At this stage growth ceased to be exponential and became linear, the principal limiting factor no longer being food resources but the surface area available for lateral expansion (i.e. the size of the beaker). As the mat of leaves became progressively thicker, the lower leaves, which were deprived of light, lost weight and eventually died. The population had now reached a stage of equilibrium in which the rate of loss of leaves equalled the number of new fronds produced.

Similar studies can be carried out with animals such as the vinegar eelworm, *Anguilla aceti*, a small nematode about two millimetres long which frequents vinegar vats. Here again, population increase follows a sigmoid curve, but the limiting factor determining the phase of equilibrium appears to be a deterioration of the environment rather than any other form of competition between individuals.

Expanding populations of *Lemna* and *Anguilla* can, in a sense, be said to illustrate comparable situations. In *Lemna*, density was self-regulating in that the rate of growth decreased as the number of plants increased. The *Anguilla* populations eventually checked their own growth by exhausting the available food supply, no doubt contaminating their environment in the process. In both, the changes observed were directly related to increasing density – they were **density-dependent**. By contrast, many plant populations are held in check by the nature

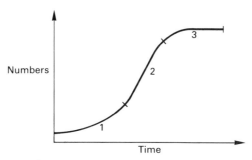

Fig. 3.2 Sigmoid curve of population growth: 1 – exponential phase; 2 – linear phase; 3 – static phase

of the physical environment, such as the drier conditions bordering a hydrosere which prevent the colonisation of land by plants growing in the water. We saw in Chapter 1 (p. 10) how a population of ragwort, *Senecio jacobaea*, which colonised an area of burnt ground was gradually overtaken during a five-year period by a rapidly expanding population of the rosebay willow herb, *Epilobium angustifolium*. This had important consequences not only for the balance of densities between the two plant populations, but also for the animals that depended upon them for food. Here, the factors controlling growth were unrelated to the density of the population concerned – they were **density-independent**. In general, density-dependent factors are those whose effects increase as the density of the population increases. Examples are food shortage, disease, and numbers of predators. Density-independent factors are unrelated to population density, and include such things as changes in the biotic or abiotic environment, and the activities of man, for example the use of pesticides.

Odum [2.6] has drawn attention to a third type of relationship between density and growth rate, that proposed by Allee in 1961. In some species, reproductive success achieves a maximum at neither a high density nor a low one, but at some intermediate value. Overcrowding, as we have seen, causes a rapid decrease in the numbers of animals or plants in a population. This may be accompanied by spectacular migrations such as occur periodically in the eastern Scandinavian lemming, *Myodes lemmus*, and the African migratory locust, *Locusta migratoria*. Undercrowding, too, has its disadvantages, as has been shown experimentally in the fruit fly, *Drosophila*, where low density is associated with a reduction in life expectancy. In sexually reproducing organisms, the smaller numbers become, the less likely the two sexes are to encounter one another since they will be dispersed over a wide area. In colonial species such as gulls, the stimulus to pair is probably induced not just by courtship between a single male and female but by the cumulative effect of the individuals in the vicinity. Below a certain critical density the birds cease to pair and a collapse of the population results. Slight crowding evidently acts as a stimulus to further increase in numbers. This may well be due to the higher level of overall activity that must result from greater competition for living space and a limited supply of food.

A schematic representation of the three patterns of growth has been devised by Odum [2.6] and this is reproduced in Fig. 3.3. The growth of *Lemna* populations

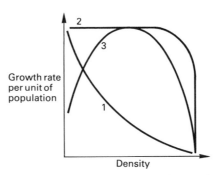

Fig. 3.3   Three patterns of population growth in relation to density: 1 – growth rate decreases as density increases (density dependent); 2 – growth rate increases until factors outside the population limit it (density independent); 3 – growth rate is highest at intermediate densities (after Odum)

referred to earlier conforms to curve 1 where the number of additions to the population per individual per unit of time decreases as density increases (**density-dependent**). Lemming and locust populations behave in accordance with curve 2 where the growth rate per individual increases rapidly and the population increases exponentially until a high density is reached and external limiting factors exert their effect (**density-independent**). Curve 3 is typified by gulls and other colonial species where growth rates are highest at intermediate densities (**intermediate density-dependent**).

Variations in density

As long ago as 1927 the German ecologist Karl Friedrich pointed out that the relationship of the environment with plant and animal communities is **holocoenotic**. That is to say, the ecosystem and the factors influencing it act as a whole. There are no barriers separating one environmental factor from another or one portion of the biotic environment from the next. This implies that patterns of response by different communities to variations in the environment are likely to be complex and difficult to interpret. This is, indeed, what we find.

The study of fluctuations in the number of animals or plants in a population owes much to the pioneer work of Charles Elton [3.2] who obtained data from the Hudson Bay Company in northern Canada on the number of skins obtained from different species of mammals (Fig. 3.4). The records show that the snow-shoe rabbit, lynx, and red fox populations all fluctuated over approximately eleven-year periods, while the oscillations of the arctic fox were three-yearly, corresponding to the lower peaks of abundance of the other species.

We now know that cyclical maxima in the size of populations are fairly common and these may be accompanied by sensational migrations such as those of

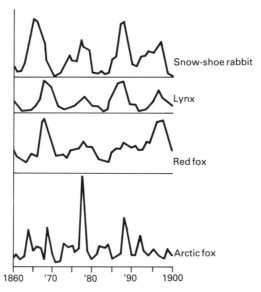

Fig. 3.4 Fluctuation in the numbers of northern mammals. The graphs show the number of skins brought in to the Hudson Bay Company from 1860–1900 (after Hewitt and Elton)

lemmings in Scandinavia (approximately a four-year periodicity) and locusts in Africa. In Britain, similar outbursts occur among a number of animal species, ranging from voles to butterflies (e.g. marsh fritillary, *Euphydryas aurinia*) to echinoderms (e.g. *Psammechinus miliaris*). However, we know comparatively little about the causes or effects of these outbreaks beyond the fact that periods of superabundance are invariably followed by intervals of rarity while the populations build up once more. This fall in numbers is only to be expected in circumstances where food must be short, disease probably widespread, and the numbers of parasites and predators on the increase.

In spite of the somewhat sensational situations quoted above, everyday experience suggests that while the densities of plant and animal populations undoubtedly vary somewhat from one season to the next, such fluctuations as we are able to detect are not necessarily cyclical or particularly drastic in their effects.

Fluctuations in plant populations

Among plant populations one of the most powerful factors influencing numbers is competition with other species for such resources as light, space, and mineral nutrients. The process of interspecific competition plays an important part in shaping the nature of communities and we will return to it again in Chapter 4.

Of the abiotic factors directly influencing land plants, one of the most vital is humidity. The destructive effect of a prolonged rainless period on small annuals and herbaceous perennials is comparatively easy to observe on any recently colonised site. If drought persists, its effects can extend to woody perennials, often with profound consequences. Thus, during the long dry period of 1979, numerous well-established beech trees (50 years old and more) growing on shallow limestone soils in Wiltshire failed to survive. The absence of their leaf canopy the following season caused a transfer of light energy to the sparse herb and shrub community below, resulting in a great outburst of growth and a remarkable change in the surrounding terrain.

Plants in general have to withstand a wide variety of attacks from herbivores and parasites. The larger and better established they are, the more likely it is that they will survive. The oak-leaf roller moth, *Tortrix viridana* (Fig. 3.5) is one of a hundred or so species that feed on the foliage of oak trees, and when present at a high density the caterpillars of the moths can cause almost complete defoliation.

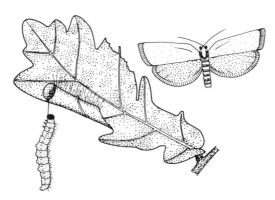

Fig. 3.5   The oak-leaf roller moth, *Tortrix viridana*, a defoliator of oak trees

(a)

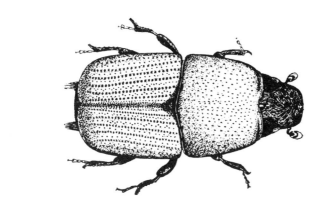

(b)

(c)

Fig. 3.6 Demise of the English elm, *Ulmus procera*, due to Dutch elm disease. (a) Left, healthy tree; centre, newly killed tree; right, two trees with extensive damage to the crown. (b) Underside of a piece of bark showing galleries made by beetles and their larvae. (c) The elm bark beetle, *Scolytus scolytus* × 20 approx. (photograph Forestry Commission)

The oak can counter this disaster by producing a second lot of leaves, but these are often badly infested by oak mildew, *Microsphaera alphitoides*, which can even lead to the death of trees in cases of severe infestation. Varley and Gradwell [3.3] measured the increase in trunk wood of oak trees at Wytham, near Oxford, and found that although the early spring growth (dependent on nutritional reserves) was not affected by heavy moth populations, defoliation could cut down summer production to two-fifths of the normal amount (judged by the size of the growth rings). It has long been observed that the nesting season of titmice and other insectivorous birds in oak woodland is closely synchronised with the invasions of these moth larvae, and it is significant that when a second crop of leaves occurs in July it is often almost totally denuded of insect life.

Whereas the oak trees under attack by defoliators and fungal parasites usually manage to survive, the same cannot be said of the elm. Dutch elm disease (Fig. 3.6) whose consequences are now so sadly widespread in Britain, was imported in logs from Holland from about 1928 onwards. It is caused by the fungus, *Ceratostomella ulmi*, which is transmitted by the elm bark beetle, *Scolytus scolytus*, a common tunneller into trees of all ages. Once infested by the fungus, the tree seldom survives, and it is no exaggeration to say that the death and subsequent removal of elms has radically altered the familiar country skyline throughout much of rural Britain. By the end of 1977, it was estimated by the Forestry Commission that of a former total population of 23 million elms, approximately 9 million had been killed by the epidemic.

Fluctuations in animal populations

Mention was made earlier of the cyclical changes in numbers that occur among populations of northern mammals and other animals. Such evidence as is available from the systematic study of plant and animal densities suggests that patterns of change exhibiting a degree of regularity may well be exceptional. Another deviation from normality is when a gradual build-up to a high density is followed by a sudden crash to rarity such as occurs in the field vole, *Microtus agrestis*. Evidently, such fluctuations are sometimes synchronised between different populations but frequently they are not. Evidence from laboratory experiments suggests that under conditions of overcrowding these small mammals undergo important physiological changes which profoundly affect their behaviour patterns. Much fighting occurs and the animals exhibit a stress syndrome (shock disease) associated with endocrine abnormalities such as enlarged adrenal glands. The general physical condition of the population deteriorates and the level of reproductive success falls. Shock disease has been reported in several species of mammals when subjected to gross overcrowding. Among invertebrates, its occurrence appears to be rare.

Parasitism as a limiting factor

In animal populations, the most usual situation is that the number of animals in the population oscillates continually to varying degrees, but with no particular periodicity. A unique record of this is provided by a colony of the scarlet tiger moth, *Panaxia dominula* (Fig. 3.7) at Cothill, Berkshire, which has had yearly estimates of its numbers made by Ford and his co-workers [2.9] since 1939. The insect is a day-flying species and this has permitted the use of the mark-release-

(a)

(b)

Fig. 3.7 The scarlet tiger moth, *Panaxia dominula*: (a) adult at rest (photograph Frank W. Lane); (b) larva feeding on stinging nettle, *Urtica dioica* (photograph S. Beaufoy)

recapture technique (described in Chapter 5 p. 94) to determine density. During the 1940s the adult population oscillated between 1000 and 8000, while in the 1950s numbers continued to fluctuate upwards reaching a maximum of 18 000. By 1961 they had begun to fall, reaching about 1400, and in the following year another reduction occurred. Subsequent years showed a gradual build-up, again accompanied by annual oscillations of varying size (see Fig. 3.8).

Although not conforming to any cyclical pattern, the changes in the density of the scarlet tiger population over the period of study were considerable. This prompts the question as to which factors were responsible for these fluctuations. Observations in the field, and laboratory experiments have shown that both larvae and adults are distasteful to their most likely predators (for example, birds) which are not, in fact, a main source of mortality. The favourite food plant of the larva is comfrey, *Symphytum officinale*, and in early summer, when they are growing rapidly, larval populations can sometimes be found at a high density. However, it does not necessarily follow that the size of the adult population is comparable, for a drastic mortality tends to occur during the last larval instar and in the pupal stage. Evidence of parasitisation by other insects is scanty and although the pupae are occasionally attacked by the wasp, *Pteromalus puparum*, the incidence is low. The main cause of mortality appears to be microorganisms – bacteria, fungi or viruses. Kettlewell [3.4] estimated the size of the larval populations in another locality near Cothill by the technique of mark-release-recapture, described in Chapter 5 (p. 94). Marking animals like larvae which periodically change their skins presents obvious difficulties, but Kettlewell had the ingenious idea of using a radioactive tracer (sulphur-35) which was supplied to the larvae in their food. By releasing a large number of marked larvae into the wild population (some 1227 were used), allowing them to randomise and then finding

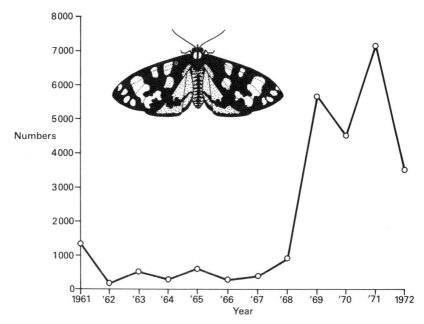

Fig. 3.8 Variation in the size of a population of the scarlet tiger moth, *Panaxia dominula*, at Cothill, Berkshire over a period of 12 years (after Ford)

the proportion of marked individuals recovered in subsequent samples (detected by means of a Geiger counter), an estimate of the total size of the larval population could be made and eventually compared with that of the emerging adults. It was found that approximately 200 000 larvae yielded about 23 000 moths, indicating a mortality of around 88 per cent. A comparable but smaller study carried out by Cook in Oxford arrived at a mean larval mortality of 94 per cent.

These studies of the scarlet tiger moth are important not only in providing quantitative evidence of the effects of density-independent factors in controlling population size, but also in showing how the magnitude of such factors can vary from one season to the next. Such variations must be regarded as the rule rather than the exception. Thus studies of parasitisation by the ichneumon wasp, *Apanteles tetricus*, of larvae of the meadow brown butterfly, *Maniola jurtina*, over a five-year period, showed oscillations in mortality ranging between 79 and 61 per cent, while mortality of pupae due to bacterial infection varied between 3 and 20 per cent [3.5].

Food as a limiting factor

There is much evidence to suggest that among animals the nature and abundance of the food available are two of the most powerful factors determining the existence and size of populations. The justifications for this statement are of two kinds, indirect and direct, and it is worth considering both of these in a little more detail.

Ever since the classic work of Eliot Howard on birds [3.6] in the 1920s, researchers have studied so-called **territorial behaviour** in a wide variety of animals, ranging from mammals, birds and fishes, to butterflies and bees. In all such species, behaviour during the breeding season assumes a characteristic pattern in which a zone (**territory**) is appropriated and defended against intruders. Territoriality is particularly well developed in birds and can easily be observed during the early months of the year. Its main features are as follows:

(i) the male leaves the winter flock and arrives in a locality a few weeks in advance of the females. He selects a prominent song post from which he sings and displays.

(ii) once he has attracted a mate, pairing occurs and the nest is built within the territory, often in the vicinity of the male's song post.

(iii) both sexes fight to maintain the boundaries of their territory against intruders of the same species, the intrusions of other species being tolerated in varying degree. Fighting is usually a ritual affair and it is seldom that the loser is seriously hurt.

(iv) the outlines of territories often follow natural boundaries such as buildings, walls and hedges, but for a particular species they vary greatly in size from year to year and from one place to another. In cliff-nesting birds such as gulls, the territory may amount to less than one square metre, while in species occupying open countryside it may extend to many hectares.

(v) territorial behaviour is not entirely confined to the breeding season; some species such as the robin, maintain a territory in the winter as well. The ecological significance of this is still uncertain.

The role of territorial behaviour in birds is no doubt partly to ensure that the male acquires a mate. Once pairing has occurred the possession of a territory probably serves the kindred function of providing a rendezvous and keeping the pair together. But besides its significance for reproduction, there are also important implications for feeding. Observations show that in many species, while foraging may range far outside the boundaries of the territory, the majority of the food is collected within it. Such behaviour may thus serve the important purpose of parcelling out the available food supply by limiting the number of breeding pairs that can colonise a particular habitat.

The operation of this principle is well illustrated by the studies of Southern and his associates on the tawny owl (*Strix aluco*) population in Wytham Wood, near Oxford [3.7]. The birds are nocturnal and exhibit marked territorial behaviour. Their call is characteristic and this can be used both to locate them and to estimate population density. Moreover, their breeding sites are easily detectable and they nest readily in suitable boxes, so breeding success can be monitored with considerable accuracy. The principal foods of the owls in Wytham were the wood mouse, *Apodemus sylvaticus*, and the bank vole, *Clethrionomys glareolus*, and the density of their populations was estimated by trapping. Southern carried out an analysis of a number of key factors relating to the dynamics of the tawny owl populations over a period of ten years. Two of these are of particular significance in relation to food supply, namely the number of eggs expected but not laid through failure of adults to breed, and the deficiency of eggs laid compared with those expected through nesting pairs failing to lay the maximum clutch size. For a particular year, when these two variables were compared with the population density of rodents in the preceding winter, it was found that losses due to failure to breed or to lay a clutch of maximum size showed an inverse relationship with the numbers of potential food animals. This became evident below a limiting value of 20 rodents per hectare, while the losses increased greatly as the availability of prey diminished. In Wytham Wood, the principal factor governing population changes was thus the level of abundance of mice and voles, which determined the breeding success of the owls. Moreover, as the number of territories increased, the average number of young fledged per pair fell from about 1.2 in 1947 to around 0.5 in 1957, both years when there was no shortage of rodents in the wood. The situation was density-dependent, breeding success declining as the number of pairs increased, accompanied by a corresponding decrease in the size of their territories. The population study illustrates an important ecological principle, namely, that among animals food provides one of the most powerful ecological factors promoting intraspecific competition. This contrasts with the usual situation among plants, where, as we have seen (p. 7), physical factors such as light and humidity play a more important role in influencing survival within a species.

Parasitism and food

A special aspect of the influence of food as a primary factor controlling populations is that of parasitism. A parasite is usually an organism which is relatively small and lives in or on its host, thus obtaining from it both a source of energy and a suitable habitat. The situation can be contrasted with that of a predator, which is free-living and usually larger than its prey, obtaining its food from it but not its habitat. The intimacy of the relationship between parasite and host varies greatly. At one extreme, the flea (an **ectoparasite**) on a dog shows a relatively loose association involving comparatively little structural specialisation by the parasite for its mode of life. On the other hand, internal parasites (**endoparasites**) such as the human tapeworm, may show extremes of specialisation in structure and mode of life. Some parasites are ectoparasitic at one stage of their life history and endoparasitic at another; others are parasitic at one time and free-living at another, and so on.

Some aspects of parasites, and ways in which they differ from free-living species are as follows:

(i) parasitism can involve no physical association with the host and may occur in the young stage only, as in the cuckoo.

(ii) some parasites (**facultative parasites**) are also capable of a free-living existence. An example is the river lamprey, *Lampetra fluviatilis*, which can be a parasite of fish but can also live independently on small fauna and fish spawn on the river bed.

(iii) the majority of endoparasites are parasitic as adults. In many, such as the roundworms (Nematoda), the young stages are free-living. In others the young stages are parasitic on other species, which may act as temporary quarters during part of the life cycle (**secondary hosts**) or merely as a means of dispersal (**vectors**). Many secondary hosts serve both functions, examples being the anophelene mosquito which transmits the parasite of human malaria (*Plasmodium*), and the water snail, *Limnaea truncatula*, which houses the larvae of the fluke, *Fasciola hepatica*, which can be a serious pest of sheep pastured in permanently wet areas.

(iv) it is to the disadvantage of a parasite to kill its host, at least until after the parasite has successfully reproduced. However, some parasitic species cause profound modification in their hosts, for example the crustacean, *Sacculina*, which inhibits moulting and causes sterility in the shore crab, *Carcinus*.

(v) in order to survive in a changing environment, it is an advantage for any organism to remain as flexible as possible both structurally and physiologically. However, a parasitic mode of life usually demands some degree of **specialisation** with a corresponding reduction in adaptability and increase in specificity where the range of hosts is concerned. The human tapeworm, *Taenia solium*, provides an extreme example. However, the parasitic roundworms (Nematoda) have achieved remarkable success as colonists of diverse habitats, ranging from the gut of man to the nephridia of earthworms, and from vinegar vats to the beer mats of pubs. They are undoubtedly some of the most ubiquitous parasites of plants and animals but appear to have undergone surprisingly little structural specialisation. Among flowering plants, dodder (*Cuscuta*) which parasitises heather and gorse, has lost its chlorophyll and relies almost entirely on its host for its resources.

(vi) specialisation in structure, physiology, and ecology leads inevitably to the need for an even more complex range of requirements so that the life cycle may be successfully completed. Hence the need for the secondary hosts and/or vectors referred to earlier. Thus another characteristic of specialised parasites is that they are not only *specific* to particular hosts, but also tend to have a *more complicated life cycle* than free-living organisms.

(vii) a vital problem facing endoparasites is that of reproduction and dispersal of young, since they are isolated inside the host. The young often escape from an animal host via the mouth or anus, or through the surface of the skin. Once liberated, the chance of the young reaching a favourable environment is frequently slight and this explains why most endoparasites exhibit enormous fertility. For instance, the human roundworm, *Ascaris lumbricoides*, is said to lay approximately 1500 eggs daily. Hermaphroditism and self-fertilisation are also common among endoparasites, both tending to increase the chance of producing large numbers of potential young.

Interactions between species

Among many plant and animal species a state of neutralism exists where populations occur independently of one another and do not interact in any way. The wider two populations are separated on the biological scale the more likely this is. As we have seen, intraspecific competition tends to be to the detriment of the species involved, a subject to which we will return in the next chapter. Parasitism and predation, on the other hand, both benefit one species while being disadvantageous to the other. Odum [2.8] has contrasted such associations with interactions of a more beneficial kind and these are summarised below.

Table 3.1   Some possible interactions between two species (after Odum)

Type of interaction	*Species* 1	2	*Nature of interaction*
Neutralism	o	o	Neither population affects the other
Competition	—	—	Direct or indirect inhibition of each species by the other
Amensalism	—	o	Species 1 inhibited, 2 not affected
Parasitism	+	—	Species 1 (parasite) generally smaller than 2 (host)
Predation	+	—	Species 1 (predator) generally larger than 2 (prey)
Commensalism	+	o	Species 1 (commensal) benefits while 2 (host) not affected
Protocooperation	+	+	Interaction favourable to both but not obligatory
Mutualism	+	+	Interaction favourable to both and obligatory

(o) no significant interaction
(+) benefit in terms of growth, survival or other attribute
(—) inhibition of population growth or other attribute

It may well be that one of the ways in which parasitism evolved was by way of **commensalism** where one partner benefits and the other remains unaffected. Thus the blind, white species of woodlouse, *Platyerthus hoffmansegii*, lives almost exclusively in the protected environment of ants nests, chiefly those of the common yellow ant, *Lasius flavus*. The woodlouse feeds on the ant faeces but whether there is a more intimate relationship between the two species is not known. A short step from commensalism is **protocooperation** when the interaction is favourable to both partners but not obligatory. Marine examples abound, for example the sea-anemone, *Calliactis parasitica*, which colonises the outside of the empty whelk shell occupied by the hermit crab, *Eupagurus bernhardus*. Presumably the crab benefits from the camouflage it obtains as well as the protection afforded by the stinging cells of the anemone. In return, the coelenterate is transported and supplied with scraps of waste food.

The condition of **mutualism** could be regarded as the ideal ecological situation, as the participant species cooperate to the mutual benefit of both. Examples are quite common, such as the much studied relationship between the freshwater coelenterate *Chlorohydra viridissima* and the unicellular green alga,

Fig. 3.9 The fly agaric, *Amanita muscaria*. The fungus forms a mycorrhizal association with the silver birch tree, *Betula pendula*

Carteria, also the corals and their zooxanthellae. Plants, too, provide some classic examples, such as the lichens which depend upon an intimate physical and physiological relationship between a fungus and a unicellular alga. The associations of legumes (family Papilionaceae) with bacteria of the genus *Rhizobium*, and between certain trees and the hyphae of fungi (mycorrhiza) are other instances of mutualism. Thus the silver birch tree, *Betula pendula*, is associated with the basidiomycete fungus, the fly agaric (*Amanita muscaria*) (Fig. 3.9) with which it forms mycorrhiza. However, its association with the bracket fungus, *Piptoporus betulinus* (Fig. 3.10) is less beneficial, for although the fungus begins as a saprophyte it frequently becomes parasitic, killing its host once its spore bodies have been produced.

Biological control of pest populations

Under natural conditions, the destruction by parasites of their host plants and animals is enormous. As we have already seen (p. 37), it is quite common among butterfly populations for around 80 per cent of their larvae to be eliminated through parasitism (mainly by hymenopterous insects such as *Apanteles*). Predation by insectivores such as birds and shrews provides a further powerful means of checking excessive increases in population density. The type of control exerted by parasitism and predation on the size of populations is sometimes referred to as **biological control**. The part it plays is significant not

Fig. 3.10 The bracket fungus, *Piptoperus betulinus*, growing on silver birch, *Betula pendula*. The association with the tree is at first saprophytic but often becomes parasitic, eventually killing the tree once the spore bodies of the fungus have been formed

only under natural conditions but also when employed as an aid to agricultural technology.

In many pests such as the large white butterfly, *Pieris brassicae* (whose larvae frequently cause havoc among *Brassica* crops (cabbages etc.) during late summer) biological control is probably largely responsible for keeping populations in check. In the large white this control is achieved by the ichneumon wasp, *Apanteles glomeratus*, which lays its eggs in the larvae during their first instar, causing afflicted individuals to die before reaching the pupal stage. The process is illustrated in Fig. 3.11.

The artificial application of biological control as a means of pest extermination has achieved its principal successes abroad. However, there are two well-known examples from Britain, both associated with crops grown in glasshouses. Thus,

(a)

(b)

Fig. 3.11 Biological control of the large white butterfly, *Pieris brassicae*: (a) adult male, (b) larvae surrounded by cocoons of the hymenopterous parasite, *Apanteles glomeratus* (photograph S. Beaufoy)

the red spider mite, *Tetrarynchus urticae*, which attacks cucumbers, has been successfully controlled by introducing the mite, *Phytoseiulus persimilis*, which feeds upon it. Here, biological control was achieved by means of a predator. Better known is the trouble caused by the greenhouse whitefly, *Trialeurodes vaporariorum*, which is a pest of tomatoes and a number of other plants grown under glass. Control was achieved by the introduction of a parasite in the form of a small chalcid wasp, *Encarsia formosa*, which lays its eggs in the bodies of the immature flies ('scales'). Those which are parasitised soon show up, as they turn black, and eventually die. Biological control by *Encarsia* was first introduced in 1926 and has since been practised on a commercial scale, the parasites being supplied to growers when they are still inside the young whitefly. Population control using *Encarsia* is highly effective, but the process is inevitably self-exhausting. Once all the pests have been killed there is no further food for the wasps and they, too, die. The glasshouses have therefore to be re-stocked with *Encarsia* each time a crop is planted.

Perhaps the most sensational example of biological control in Britain was of the rabbit population by myxomatosis in 1953, which is estimated to have reduced the numbers by more than 90 per cent. The myxoma virus is indigenous in South America where it is not particularly virulent. When populations are infected there, the parasite is passed from one host to another by means of the mosquito (in Europe the vector is the rabbit flea). After several attempts, myxomatosis was successfully introduced into Australia in 1950, from there it was brought to France and thence to Britain. The sudden disappearance of rabbits had extraordinary and sometimes drastic consequences. Carnivorous birds such as the buzzard failed to breed for two years until they found alternative food sources. Land such as chalk downland which, traditionally, had been heavily grazed, was suddenly faced with a reduction in grazing pressure. As a result, plants like perennial grasses which had previously been kept in check, flourished, and the chalkland assumed the appearance of an uncut hayfield. Many of the characteristic chalkland flowering plants were stifled and died, some of the rarer ones such as the rampion, *Phyteuma orbiculare*, becoming extinct in areas where they had previously been common.

As Fletcher has pointed out [3.8], there are a number of appealing features of biological control when employed on a commercial scale:

(i) used systematically it has proved highly efficient. More than 100 different species of pests have been controlled in this way in some 60 different countries. Moreover, it has worked equally well in a diversity of climates and ecological conditions. There are some who believe the development of the technique is only in its infancy, and that its more widespread application is only a matter of time and knowledge. As we shall see, this view is open to question.

(ii) one of the criticisms of chemical pesticides is that they frequently exterminate a variety of other species as well as the pest, some of them being beneficial. Also, there are several instances known where the indiscriminate killing of one insect pest has resulted in the upsurge of another. Parasites and predators used in biological control are always highly specific in their effects.

(iii) whereas many chemical pesticides leave behind them undesirable residues, biological control is clean and does not cause contamination of any kind.

(iv) compared with chemical pesticides, biological control is relatively cheap.

While much work may be entailed during the initial stages of isolating and culturing a predator or parasite, little expenditure is needed thereafter. Calculations by the Commonwealth Institute of Biological Control [3.8] have shown that for seven successful biological control ventures, an overall expenditure of £1 000 000 yielded aggregate benefits totalling £4 750 000 and subsequent annual savings of £262 000.

Integrated control of pests

In recent years, just as the development of methods for the biological control of pest populations has made great strides, so too, has the production of more effective synthetic pesticides. In general, the trend has been towards finding substances which are more specific in their action, which do not contaminate food crops and which leave behind them no serious residues in the soil. One of the reasons why the control of whitefly in glasshouses by *Encarsia* has proved so popular is that no insecticide was known which killed both the adults and the young stages (including the eggs). In order to eradicate the pest, a programme of repeated spraying was necessary to exterminate each new batch of adults as they appeared. But the advent of modern pesticides such as permethrin which is effective against eggs, young and adults, has radically changed the situation. Another problem has been that the persistent use of a particular type of spray has resulted in the evolution of resistant strains of the pests. Thus, in many glasshouses there are now whitefly and greenfly populations that are unaffected by the widely used insecticide malathion.

There is obviously a strong argument in favour of a more coordinated approach to pest control which combines the best features of biological and chemical methods, while causing the minimum disruption of closely related but beneficial ecosystems. Such programmes are often referred to as **integrated control**. Their aim is the preservation of useful natural enemies and the destruction of the pests themselves. One approach to the problem is to ensure that the pest and its natural predator are not equally exposed to a particular pesticide. For instance, it may well be possible to build up reservoirs of predators by leaving adjacent fields untreated. Again, populations of predators could be bred in the laboratory and subsequently released, as has been done with great effect against the scale insects infecting Californian citrus trees.

Possibilities for the future

In the future it may be possible to exploit differences in physiology and behaviour of pest and predator. For instance, the specialised feeding habits of many predators and parasites could provide them with immunity from chemical treatments. Endoparasites will be protected from contact sprays although they may subsequently succumb if the pest itself is seriously contaminated. With pesticides there is a tendency to use them at a concentration above that really needed for a particular kind of control ('overkill'). Reducing dosage can have a selective effect, enabling predator reservoirs to survive, and if the pest population is not totally eliminated, allowing prompt recolonisation by predatory species.

In conclusion we may say that while biological control methods have undoubtedly achieved considerable success, their principal limitation lies in the identification and subsequent culture of suitable predators or parasites which will not only eliminate the pest in question but disregard related beneficial species to which they might also prove partial. Compared with the total range of problems in pest control that now confronts us, the number of predators likely to be available is quite small. For this reason, it seems that biological control is unlikely ever to be the main weapon of attack on pests. Its undoubted benefits compared with chemical pesticides will ensure its continued use and development, but the future of pest control must lie in integrated schemes which exploit to the full the respective advantages of chemical and biological methods.

Control of human populations

From the previous pages it will be seen that, under natural conditions, the density of plant and animal populations is in a constant state of change. Sometimes these changes follow a cyclical pattern; more often they do not. The interactions of organisms both with one another and with their physical environment are exceedingly complex. Nonetheless, when suitable techniques and approaches are devised, it is sometimes possible to analyse the nature of population changes with some precision and to identify their underlying causes.

Among living organisms man is unique in that he is potentially capable of regulating his own numbers. In contrast to what has been said so far, the control of human populations is therefore an appropriate note on which to end this chapter. According to agencies of the United Nations, whereas the world population is now about 4400 million, its estimated size in year 2000 is around 6400 million. The contributions likely to be made to this increase by the developed countries of the west and the developing countries provide a striking contrast. Whereas the annual rate of increase of developed countries in 2000 is projected as 0.6 (now 0.9), that of the developing world is 2.2 (now 2.4). It is also worth noting that although the annual death rate in developed countries is likely to remain unchanged at about 10 per 1000, that of the developing countries is forecast to fall from its present value of around 14 per 1000 to 8 per 1000 in 2000.

As Odum [2.6] has pointed out, variations in human culture are usually adaptive in that they have evolved along with man himself. However, when circumstances change, behaviour patterns which were formally advantageous can become a disadvantage. A high rate of reproduction, for instance, is beneficial when density is low and a surplus of resources is available, but it is disadvantageous when density is high and resources are becoming depleted. As the figures quoted above clearly illustrate, during periods of rapid human population increase the birth rate remains high but the death rate drops as we see in the developing countries. Once the standard of living starts to rise the birth rate begins to decline as has happened in the developed world. Thus, one of the most powerful tools available for controlling population density is adjustment of the standard of living. But as we see in parts of Africa and elsewhere, where sources of energy and other materials are limited, rapid population growth inhibits the economic and social development that might otherwise raise the standard of living. Hence the reason for the extreme poverty and continuing high birth rate in these regions.

It therefore follows that to rely on raising the standard of living as the only means of population control is not enough. Other more direct forms of self-regulation are now needed. Some of these are medical, such as programmes for birth control and even voluntary sterilisation. The likelihood of their success depends largely on more effective education and changing long established attitudes. Other approaches may be social, such as changing the status of women so that they have equality with men in decision-making on such matters as family planning. Again, control may be political and economic through tax incentives for reduced family size and penalties for too many children.

As we have seen earlier (p. 34), excessively high population density brings with it a number of characteristic symptoms in addition to mass starvation, such as the stress syndrome typified by increased aggression. It may be that we are already witnessing the beginning of such effects among the poorer areas of South America. The choice for human survival is either to reduce the birth rate or to suffer the effects of an increased death rate. World-wide disease, starvation, pollution, war, and social upheaval are none of them attractive alternatives.

A particular problem facing the world human population is the availability of food. The Food and Agriculture Organisation (FAO) of the United Nations has estimated that by the year 2000 there will be a world short-fall of some 20 million tonnes of protein. Here the outlook is more optimistic, for it seems likely that improvements in agricultural methods can make a substantial contribution to closing the gap. The problem is largely economic: that of finding the means of transferring food from areas of high production to those where populations are less favoured.

Summary

1 The size of a population depends on a balance of reproductive success and death rate. Its numbers are expressed as its density.

2 The pattern of population growth conforms to an S-shaped curve and passes through three distinct phases.

3 The factors influencing population growth can be dependent or independent of population density. Many examples have been studied which illustrate the mode of action of these factors in different circumstances.

4 Fluctuations in population density sometimes occur in a regular cyclical manner. More often, variations in numbers are erratic and difficult to predict.

5 Parasitism can be a significant factor influencing density. Its effects are often particularly severe during the early stages of the life cycle in both plant and animal hosts.

6 Food availability is another important density-dependent factor. Among animals such as birds, territorial behaviour is important in parcelling out a limited food supply.

7 Parasites are faced with peculiar problems in obtaining food from their hosts and possess a variety of adaptations which aid their survival.

8 Interactions between two species can assume a variety of forms ranging from neutralism where neither population benefits, to mutualism where the relationship is favourable to both.

9 Biological control involving the destruction of pests by the introduction of their parasites or predators has proved successful in a number of instances but has considerable limitations.

10 Integrated pest control relies on the balanced use of parasites or predators and chemical pesticides so as to limit upset of the environment to a minimum.

11 Ecological principles can be useful in pointing the way towards methods of controlling human populations more effectively.

References

3.1 Clatworthy, J. N. and Harper, J. L. (1962) 'The comparative biology of closely related species living in the same area. *V* Inter- and intra-specific interference within cultures of *Lemna* spp and *Salvinia nutans*', *Journal of Experimental Botany*, **13**, 307–24

3.2 Elton, C. S. (1924) 'Periodic fluctuations in the numbers of animals; their causes and effects', *British Journal of Experimental Biology*, **2**, 119

3.3 Varley, G. C. and Gradwell, G. R. (1962) 'The effect of partial defoliation by caterpillars on the timber production of oak trees in England', *Proceedings of the 11th Congress of Entomology*, **2**, 211–14

3.4 Kettlewell, H. B. D. (1952) 'Use of radioactive tracer in the study of insect populations (Lepidoptera)', *Nature*, **170**, 584

3.5 Dowdeswell, W. H. (1961) 'Experimental studies on natural selection in the butterfly, *Maniola jurtina*', *Heredity*, **16**, 39–52

3.6 Howard, H. E. (1948) *Territory in Bird Life*, Witherby

3.7 Southern, H. N. (1970) 'The natural control of a population of tawny owls (*Strix aluco*)', *Journal of Zoology*, **162**, 197–285

3.8 Fletcher, W. W. (1974) *The Pest War*, Blackwell

4

Communities and ecosystems

In the previous chapter we saw how populations, although superficially simple units in that each consists of a single species, are, in fact, highly complex. Interrelations between individuals are not only internal (intraspecific) but also extend to other populations such as parasites, competitors, and prey (interspecific). The theme of the present chapter is the community: a group of populations occupying a particular kind of habitat.

Balance of numbers

Prolonged studies of natural plant and animal populations have shown that while their density fluctuates somewhat from year to year, these fluctuations are usually within fairly narrow limits and occur in an erratic manner. Occasionally, a catastrophe such as the advent of myxomatosis in rabbits or dutch elm disease (p. 34) occurs, from which a population may not recover. However, such extreme occurrences are comparatively rare. In the absence of human interference, the general impression that one gains from the study of habitats such as ponds, streams, and woods is that the species present and their frequency remain fairly stable from year to year. Although communities involve complex and changing interactions, they nonetheless exhibit a high degree of stability (*homeostasis*). In this sense they are comparable to the systems which control, say, mammalian body temperature or balance. An essential feature of all homeostatic mechanisms is that they serve to correct any tendencies in dynamic systems to move away from a steady state or norm in one direction or another. A much oversimplified example is given in Fig. 4.1 which illustrates the influence of positive and negative feedback systems in controlling the numerical balance between an animal population and its predators.

Increasing numbers of potential prey will provide resources for a greater number of predators, but as the predators increase so will the pressure on the prey population which will therefore tend to decline. Although this example illustrates

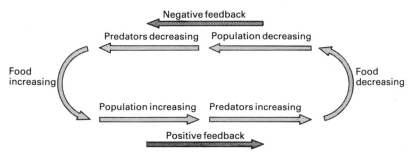

Fig. 4.1 Simplified homeostatic system controlling the density of an animal population in relation to its predators

satisfactorily an important ecological principle of density control, it is, nonetheless, a gross oversimplification of reality in that it disregards other groups within the community with which the populations in question will also interact, such as parasites. Thus increases in the number of both plant and animal species are frequently accompanied by corresponding increases in the incidence of disease which, in terms of destruction, can exert a far more powerful influence on density than predators.

Production and consumption

Natural communities of plants and animals, in common with human communities, are subject to the basic law of economics that production must be equal to or exceed consumption. Consumers cannot survive in isolation from producers. In nearly all communities, the primary production of resources is carried out by green plants (**autotrophs**) through the process of photosynthesis. Plants provide food for herbivorous animals (**primary consumers**), and some of these are eaten by carnivores (**secondary consumers**). On comparatively rare occasions, small carnivores may provide food for larger carnivores (**tertiary consumers**). We thus have the food chain (see p. 3):

Plant → Herbivore → Carnivore
(autotroph)   (heterotrophs)

This can be expanded to:

Primary producer → Primary consumer → Secondary and Tertiary
(plant)   (herbivore)   consumers (carnivores)

For effective growth (production), some of the essential requirements of plants are a supply of water and the appropriate minerals. However, for photosynthesis to take place and hence for all life to function, a source of energy is also needed. This is supplied by the sun. Research has shown that of the huge quantity of energy reaching the earth's atmosphere (about $63.8 \times 10^3\,\mathrm{kJ\,m^{-2}\,year^{-1}}$) only about two per cent is used by plants for photosynthesis and transformed into tissues of all kinds. Of the remaining 98 per cent reaching plants, a small proportion (one per cent or less) is expended in respiration and the remainder is lost by latent heat of evaporation, reflection, and passing through the tissues unchanged.

As an outcome of the cycle of growth and death occurring among living things, we might expect a gradual accumulation of debris to occur. However, with the exception of the formation of peat (semi-decomposed plant material) this does not happen, and decomposers such as fungi and bacteria ensure that dead remains are quickly broken down and recycled by way of the nitrogen, carbon, and other cycles. This implies that a dynamic equilibrium must exist between the overall primary production of autotrophs and the level of consumption by heterotrophs. But where the transfer of energy is concerned, the utilisation of food by herbivores and carnivores is highly inefficient. Among herbivores, as much as 90 per cent of the food consumed may pass out of the body as faeces. Indeed, in large species such as the elephant, inspection of undigested remains strewn on the ground can provide a precise idea of both the qualitative and quantitative nature of the animals' diet. Carnivore digestion is more efficient and an average of about 45 per cent assimilation is normal (55 per cent wastage), although it can attain 75 per cent at the highest level. Since much of the world's protein supply depends upon the consumption of herbivores, these figures for the utilisation of energy are of great international significance and will be discussed further later (p. 282).

Energy conversion

From what has been said so far regarding the energy relationships existing at the different levels of an ecosystem (**trophic levels**), we can deduce certain characteristic attributes that are common to all communities. All ecosystems are subject to the **First Law of Thermodynamics**. This states that whereas energy may be converted from one form into another, it can neither be created nor destroyed. In other words, the total energy intake through primary producers must be accounted for either through assimilation by primary and secondary consumers or dissipation through waste products and losses to the atmosphere through respiration. Thus, adopting the schema of Phillipson [4.1]:

For autotrophs:

$$\frac{\text{Solar energy}}{\text{assimilated}} = \frac{\text{Chemical energy}}{\text{of growth}} + \frac{\text{Heat energy}}{\text{of respiration}}$$

For heterotrophs:

$$\text{(i)} \quad \frac{\text{Chemical energy}}{\text{eaten}} = \frac{\text{Chemical energy}}{\text{assimilated}} + \frac{\text{Chemical energy}}{\text{of waste}}$$

$$\text{(ii)} \quad \frac{\text{Chemical energy}}{\text{assimilated}} = \frac{\text{Chemical energy}}{\text{of growth}} + \frac{\text{Heat energy}}{\text{of respiration}}$$

From the above it follows that, bearing in mind the energy lost in the transformation that takes place at each trophic level, the biomass of producers must always be greater than that of primary consumers, while this in turn will invariably exceed that of secondary consumers.

These two general principles have important implications for the food relationships of the plants and animals comprising any community and they will be considered further in the next section.

Food relationships

The idea of a food chain is relatively simple, but when we look at food relationships in the context of a whole community (or even part of one) the situation becomes more complicated. Fig. 4.2 illustrates some of the food relationships existing within a typical pond ecosystem and these have been arranged so as to conform to the different trophic levels represented. Certain generalisations can be drawn from this diagram that apply to plant and animal communities in general.

(i) The pattern of food links is exceedingly complex. Instead of linear chains, we find a criss-cross network of relationships which is appropriately referred to as a **food web**. The main reason for its complexity is that animals consume a variety of different foods. Neal [4.2] collected data on the food of badgers, *Meles meles*, from 12 different observers of sets and his results are summarised in Table 4.1.

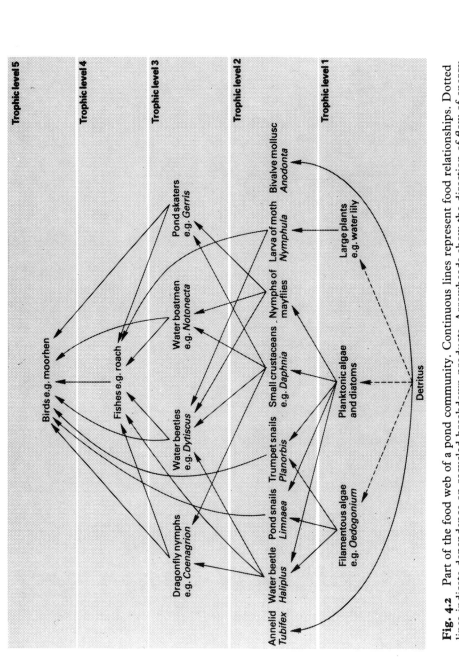

Fig. 4.2 Part of the food web of a pond community. Continuous lines represent food relationships. Dotted lines indicate dependence on recycled breakdown products. Arrowheads show the direction of flow of energy

Table 4.1   Animals and plants recorded as eaten by badgers (*Meles meles*)
(Data from Neal)

Commonly eaten		Occasionally eaten	
Animal	*Plant*	*Animal*	*Plant*
Young rabbits	Acorns	Old rabbits	Blackberries
Rats	Beech nuts	Voles	Apples
Mice	Bluebell bulbs	Hedgehogs	Corn
Slugs	Lords and ladies roots	Moles	Pig nuts
Bees and larvae	Grass	Lambs (dead)	Wild parsley roots
Wasps		Lambs (live)	Dog's mercury roots
Beetles		Birds (adult)	
Worms		Birds eggs	
		Frogs	
		Toads	
		Snails	
		Woodlice	
		Lepidopterous larvae	
		Other larvae	

It will be seen that the animal's diet is extraordinarily varied, and although primarily a carnivore, it eats a considerable amount of plant material. Moreover, its menu changes with the seasons as the different items become available. As an omnivore, it is thus extracting energy from the ecosystem at two different trophic levels. The picture is further complicated by the fact that the food of animals usually changes as they get older, and a predator on a species at one stage may become its prey at another. Thus the young fry of the trout provide a significant component of the diet of carnivorous arthropods such as water beetles and dragonfly nymphs, but once the fishes are fully grown the tables are turned.

(ii) At the bottom of the food web are the partially recycled remains of dead plants and animals (**detritus**). This provides the food, partly or completely, for a host of small animals such as annelids, molluscs, the nymphs of mayflies, and small crustaceans such as the freshwater shrimp, *Gammarus pulex*. Detritus also provides the rooting medium for larger plants such as water lilies. Eventually it is broken down by bacteria of the nitrogen cycle and the resulting minerals are released into the water, where they promote the growth of small plants such as algae. As will be seen from Fig. 4.2, these occupy a central position in the food web, although their precise relationship with herbivores at the next trophic level is not easy to represent in a diagram of this kind. Thus it will be seen that while pond snails (*Limnaea*) and trumpet snails (*Planorbis*) feed on filamentous and planktonic algae, they do not eat the leaves of the larger plants such as the water lily. Yet, common experience shows that gastropod snails of several kinds are frequently to be found on the floating leaves of water lilies and pond weeds such as *Potamogeton*. Observation in the laboratory indicates that while the snails are undoubtedly browsing on the under surfaces of the leaves, they are not eating them, but are rasping with their radulas at the film of algae that adheres to them. In such an association between two different species neither appears to benefit, and the partner which inhabits the surface of the other is known as an **epibiont**. A plant living on another plant such as the alga inhabiting a water lily leaf or the moss on the bark of trees, is often referred to as an **epiphyte**. An animal living on another animal is known as an **epizoite**, typified by the numerous stalked ciliate protozoans (e.g. *Carchesium*) that inhabit the appendages of the freshwater shrimp, *Gammarus pulex*, a common inhabitant of ponds and streams.

(iii) Because of their autotrophic method of feeding, green plants form the base of the food web. One of the reasons why ponds and shallow seas can support such a rich community of organisms is that the water is usually well illuminated and contains an abundance of minerals. Moreover, the absence of a current permits the growth of dense populations of planktonic plants and animals. In streams and rivers much of this section of the food web is washed away and the resulting community is thus poorer both in numbers and diversity of species. In areas shaded by trees and bushes the growth of green plants will be inhibited and primary production correspondingly reduced. However, autumn leaf fall results in the addition of large quantities of vegetable matter (and therefore of energy) which is soon broken down to detritus. It is quite common in woodland ponds and streams for almost all the energy required to support the animal community (often a high density of detritus feeders) to be obtained from outside.

(iv) As we ascend the food web, the number of organisms decreases but their size increases. The availability of energy within an ecosystem is such that it can only support relatively few large herbivores and carnivores. As Paul Colinvaux [4.3] has pointed out so vividly, this explains why big, fierce animals are rare. Thus, the maximum number of levels in a food web is normally five, the species occupying the highest trophic level being virtually immune to predation. But the average number of links in a chain is usually nearer three.

Ecological pyramids

In Africa, it has been estimated that the support of a single lion necessitates the slaughter of around fifty zebras or their equivalent a year. The need to quantify the relationship between predators and their prey led Elton [1.1] to coin the phrase **pyramid of numbers**. Thus, in a free living community, the different trophic levels can be represented diagrammatically, with the primary producers and consumers (consisting of many small plants and numerous herbivores) forming the base of the pyramid. As we approach the apex, numbers become smaller, size increases and predators become fewer (Fig. 4.3). Another way of presenting the same situation is shown in Fig. 4.4(a). However, the plants supporting a particular ecosystem are not always small and numerous but may be few and large, as in the case of a group of oak trees. The pyramid of numbers will then be as shown in Fig. 4.4(b). Incidentally, it is worth comparing the numerical relationships in parasite chains with those of free-living forms. In the former, the pyramid of numbers is inverted (Fig. 4.4(c)) with the apex downwards. Starting with a single organism (the host), the parasites become progressively smaller and more numerous at each trophic level.

The pyramid of numbers enables us to ascertain the numerical relationships between producers and consumers *within a particular ecosystem*. Its principal

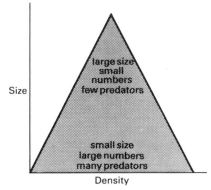

Fig. 4.3 The pyramid of numbers in a free-living community

limitation is that it provides no basis for a comparison of different ecosystems, for example a pond and a stream, where, as we have seen, the situations are quite different. The problem can be overcome to some extent if we use as our parameter the dry mass (biomass) of organisms rather than their numbers (**pyramid of biomass**). We can then determine the mass of producers required to support a given mass of consumers and this can be used as a valid basis for comparison. An apparent paradox is provided by Fig. 4.4(d), where the bulk of phytoplankton appears to be smaller than that of the zooplankton it supports. Such a result is typical of samples obtained in well illuminated static waters such as lakes and shallow seas at certain times of the year. These merely record the mass of living material present over a short period (**standing crop**). In this instance, the growth of the phytoplankton happened to be in a phase of decline while the high density of zooplankton indicated the outcome of a previous burst in plant multiplication. A

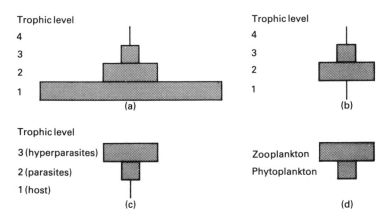

Fig. 4.4 Some ecological pyramids: (a) numbers with many small primary producers, (b) numbers with few large primary producers, (c) parasite chain of numbers, (d) inverted pyramid of biomass. Trophic levels are 1 (primary producers); 2 (primary consumers); 3 (secondary consumers); 4 (tertiary consumers)

few weeks later the situation could have been reversed and this would certainly have been so over the longer period of a year. The limitations of the pyramid of biomass, then, are that in making measurements of the standing crop, no account is taken of the total amount of material involved or of the rate at which it is produced. Plants reproduce at vastly different rates. Thus, while a green alga may multiply every few hours, many large trees produce no seed for the first five years or more of their life. In comparing the standing crops of algae and trees we must therefore take account of the rate of production (**productivity**) that has given rise to the existing situation. It is this that estimates of the standing crop fail to do.

The rate of production of living things is a function of their energy uptake. In green plants, the total amount of light fixed through photosynthesis in a given period of time is known as the **gross production** (G). Of this amount, a portion will be required for the maintenance of the life of the plant through respiration (R). The remainder is theoretically available for growth, and represents the **net production** (N). Thus:

$$G = R + N$$

To follow the transfer of energy through an ecosystem we need to know what proportion of net production (N) results in the formation of new tissues and an increase in biomass (T) and how much is lost through consumption by herbivores (C) or through death of organs or individuals (D). Thus:

$$N = T + C + D$$

For food crops, T is the crucial figure, for this indicates the amount that will be available for harvesting.

For different ecosystems the relationship between these variables fluctuates greatly, as is illustrated by a comparison of N and C. Thus in a beech forest in Germany, net primary production was estimated at $170 \times 10^4 \, \text{J m}^{-2} \, \text{year}^{-1}$, of which $6 \times 10^4 \, \text{J m}^{-2} \, \text{year}^{-1}$ (3.5 per cent) was removed by leaf-eating insects, while in a river in Florida, out of a production of $3696 \times 10^4 \, \text{J m}^{-2} \, \text{year}^{-1}$, some $1409 \times 10^4 \, \text{J m}^{-2} \, \text{year}^{-1}$ (38.1 per cent) was consumed by small crustaceans. These figures provide an interesting contrast with the situation in European agricultural grassland, where it has been estimated that in an area grazed by sheep at a density of one animal per 0.4 ha, out of a net production of $956 \times 10^4 \, \text{J m}^{-2} \, \text{year}^{-1}$, some $257 \times 10^4 \, \text{J m}^{-2} \, \text{year}^{-1}$ (25.8 per cent) were consumed.

It may well be questioned whether grassland, no matter how fertile, is capable of supporting such high pressure grazing for long without incurring a serious decline in productivity. The feeding mechanism of sheep poses a particular problem, since they pluck the grass stems, which tends to damage the roots and, over a protracted period, brings about a deterioration in the pasture. By contrast, cows (like many other herbivorous mammals such as rabbits) feed by cutting the grass blades and stems. This serves to prevent the formation of flowers and to maintain the plants in the vegetative growth phase. As we have seen, herbivores are inefficient converters of energy, which means that more than half the vegetation consumed will be returned to the ground in the form of faeces and urine which are important nitrogenous fertilisers. So, provided grazing is not excessive and is restricted to certain times of the year when the various species of grasses are growing, it is likely to result in an increase in pasture productivity.

Energy and ecosystems

Representing the transfer of energy between the different trophic levels of an ecosystem in the form of a pyramid is a rather crude procedure in that it tells us nothing of the partition of energy between the species forming the community. Moreover, as we have seen already, many predators consume a variety of foods, some of it plant and some animal, which they extract from different levels of the food web. In order to obtain a realistic picture of what is happening, we need to adopt a more precise method of recording energy flow.

If we wish to construct a flow diagram such as that shown in Fig. 4.5, two essential pieces of information are required. We must be able to measure the amount of light energy entering the system over a given period of time (usually expressed as $kJ\ m^{-2}\ year^{-1}$). This is measured by means of one or more integrating light meters (solarimeters) which register a cumulative record of the incident illumination (see Chapter 6 p. 108). We also need to know the average energy value for the tissues of each of the different species that belong to the ecosystem. This is usually expressed in $kJ\ g^{-1}$ dry mass and is obtained by burning known amounts in oxygen in a bomb calorimeter and measuring the heat

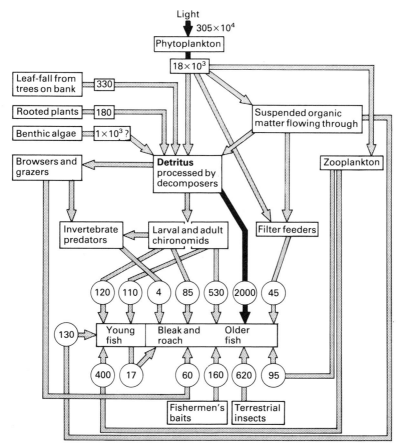

Fig. 4.5 Simplified energy flow diagram based on the River Thames at Reading. Figures are $kJ\ m^{-2}\ year^{-1}$

evolved. Thus, for the small crustacean *Calanus*, an important component of the marine zooplankton (p. 64), the energy value is approximately 30.96 kJ g^{-1}. If we can estimate the mass of *Calanus* eaten by different predators (secondary consumers) in a known period of time, we can determine the contribution of this particular food source to the total energy budget.

Berrie [4.4] studied a section of the River Thames about 4 km long in the vicinity of Reading, where the water is turbid, slowly flowing, and eutrophic (with a high mineral content). Data were obtained for the phytoplankton, benthic algae (inhabiting the river bed), rooted plants, leaf-fall from trees overhanging the river, benthic animals such as chironomids and large freshwater mussels, zooplankton, and fish. Using this information, Berrie compiled an energy flow diagram, a simplified version of which is reproduced in Fig. 4.5. Two features of particular interest are the important role played by detritus and detritivores, and the position of fish in the food web (bleak and roach). Analysis of stomach contents showed that adult bleak were predominantly carnivorous, roughly 75 per cent of their food being other animals. By contrast, only about 18 per cent of the food of older roach was aquatic animals, the remainder being detritus, algae, and insects taken at the surface, which had entered the water from another (terrestrial) ecosystem. The diet of young roach evidently differed appreciably from that of adults, being mainly carnivorous with zooplankton and chironomid larvae providing the principal items of food.

Studies of this kind are of considerable academic value in providing a precise and quantitative idea of the relationships existing within a food web and the ways in which the energy entering an ecosystem is utilised and transmitted from one trophic level to another. But they can also have an important practical significance. In an ecosystem which has become reasonably stabilised, it enables us to make predictions regarding the likely outcomes of change, for instance, if it were desired to raise the productivity of a river so as to increase the existing fish populations or to accommodate a new one.

Cycling of nutrients

The energy entering an ecosystem passes through it from one trophic level to another, and is eventually lost to the outer atmosphere in the form of heat. On the other hand, nutrients, such as water, carbon dioxide, nitrogen, and a wide variety of minerals, are absorbed either from the atmosphere or the soil, circulate through the food chains and are released again to the environment whence they came. The availability of these nutrients is finite and the process of recycling is thus one of the fundamental requirements for continuing life. Moreover, as we shall see in Chapter 14, it is a process that frequently competes with the interests of man.

Cycling takes place at two different levels: that which involves the total atmosphere, and is therefore global, and at a more localised level where only the earth's surface is involved (sedimentary).

Global cycles (atmospheric)

One of the most important of these is the water cycle (**hydrological cycle**) which is summarised in simplified form in Fig. 4.6. Particularly significant are the sources of energy that drive the system of circulation. Thus, radiant heat from the sun causes evaporation of water from the sea, forming clouds, while drainage of

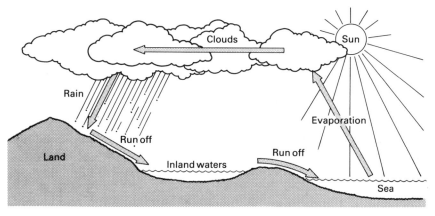

Fig. 4.6 The hydrological cycle

water from land by rivers and other natural waterways, ensures its eventual return to the oceans.

The **carbon cycle** (Fig. 4.7) is more complex, in that the eventual retrieval of the carbon derived from atmospheric carbon dioxide and converted by autotrophs into plant tissues, involves breakdown by chemical processes. The decomposition of dead plant and animal remains is achieved by microorganisms in the soil and water, mainly bacteria and fungi. Aiding this process at the halfway stage is a vast host of small animals (mostly arthropods) which live on semi-decomposed organic

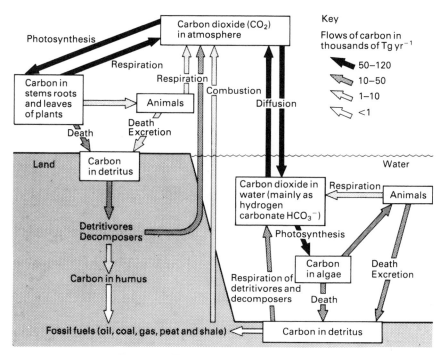

Fig. 4.7 The carbon cycle (after King)

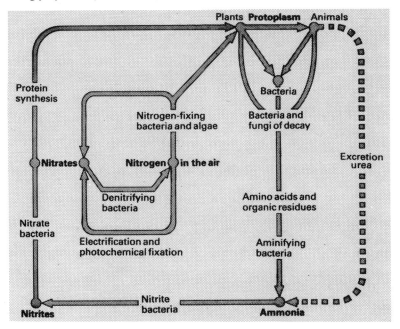

Fig. 4.8 The nitrogen cycle

material (**detritus**) and are hence known as **detritivores**. The burning of fossil fuels such as coal and oil tends to increase the carbon dioxide in the atmosphere, but this is offset, at least in part, by the effect of rain, in which the gas dissolves to form hydrogencarbonate (HCO_3^-) ions.

The **nitrogen cycle** is again ultimately dependent on the earth's atmosphere, of which the gas forms some 79 per cent (Fig. 4.8). Plants absorb nitrogen mainly

Fig. 4.9 Nodules on the roots of a leguminous plant (broad bean) housing symbiotic nitrogen-fixing bacteria of the genus *Rhizobium*

in the form of nitrates, which are highly soluble in water; they are converted into amino acids and then into proteins. Whether the plants are eaten by animals or not, the nitrogen eventually reaches the detritus stage, where it provides the food for detritivores and decomposers. The nitrogen is finally released as ammonium ions, many of which are converted by soil bacteria into nitrate (**nitrification**). Under anaerobic conditions such as in deep mud, some of these ions are converted into toxic nitrite, which is only slowly changed to nitrate provided sufficient oxygen is available. Many bacteria and blue-green algae are able to convert atmospheric nitrogen into simple inorganic compounds, a process known as **nitrogen fixation**. Best known among these are bacteria of the genus *Rhizobium*, which form a symbiotic association with many species of leguminous plants such as clover. The microorganisms are concentrated in nodules (Fig. 4.9) which represent the reaction of the host plant to the invading symbionts. Set against the gains in the form of nitrate are the losses occurring in poorly aerated soils through denitrifying bacteria. These break down inorganic nitrogen compounds and release nitrogen into the atmosphere.

Localised cycles (sedimentary)

Many cycles, such as those which achieve the conservation of vital minerals, are of a much more specific kind and are independent of global influences. An example is sulphur, which is a frequent constituent of the amino acids which form proteins. It is transmitted in aquatic ecosystems mainly in the form of sulphate (SO_4^{2-}) ions (Fig. 4.10). Much of this is utilised in primary production, some being returned to the water through the excretion of animals. Sulphur compounds remaining in dead organisms are decomposed by microorganisms to form hydrogen sulphide (H_2S) which is eventually oxidised by sulphur bacteria (**chemosynthetic organisms**) to form sulphate (SO_4^{2-}) ions.

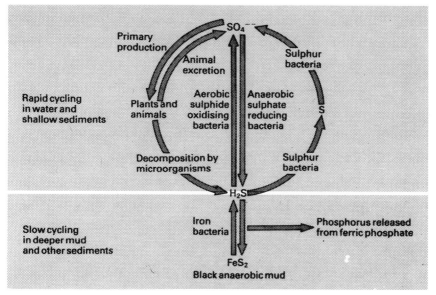

Fig. 4.10   The sulphur cycle in aquatic ecosystems (modified after Odum)

In anaerobic conditions, much of the hydrogen sulphide is not oxidised but passes into a sulphur reserve, where it combines with iron under the influence of iron bacteria and eventually forms iron sulphide (FeS_2). Unlike the aerobic cycle which takes place quite quickly in water and shallow sediments, the anaerobic phase is much slower. The various sulphur compounds formed are evil-smelling, and are characteristic of the glutinous mud frequently found beneath areas of stagnant or static water. They are all too common in poorly attended aquaria.

The different cyclic processes do not operate in isolation but interact with each other. This is illustrated by the fact that when iron sulphide compounds are formed, phosphorus is converted from an insoluble to a soluble form, thus becoming available for uptake within the aquatic ecosystem.

The sulphur cycle in soil is similar to that in water, with the changes taking place rapidly in the well-aerated topsoil and more slowly in the less aerobic subsoil deeper down.

Ecological habitats

In the preceding discussion, plants and animals were grouped together according to the trophic level to which they belong within the ecosystem. This is a useful means of classification in the construction of food webs and is a preliminary step towards a more detailed analysis of the flow of energy.

However, there is also a need to describe a community in terms of the activities of the species that compose it, in order to gain a clearer idea of how the available resources are being parcelled out between different and sometimes competing populations. This is particularly desirable when successional changes are being studied and an attempt made to predict their likely outcome (p. 67). In order to describe a community in economic terms, we have to fit the animals into the total ecological picture largely on the basis of their feeding habits. Thus, in woodland, we could classify the different insects feeding on plant leaves as leaf miners, sap feeders, gall formers, and defoliators. Again, the birds could be grouped as fruit and seed eaters, insectivores and larger carnivores.

Elton [1.1] referred to each economic unit as a **niche**, a term still used today. Here, the word **habitat** is assumed to cover not only where an organism lives but also what it does. The classification of habitats serves to highlight the availability of particular kinds of food and facilitates an assessment of the amount of competition likely to occur in particular circumstances between species with the same or similar requirements.

Practical aspects of food webs

It was mentioned earlier (p. 58) when considering food webs and the flow of energy in ecosystems, that while such studies have much academic value, their findings can also be put to considerable practical use. Indeed, this area of ecology has now become an important and expanding part of biotechnology. Two examples will illustrate the kind of contributions that such studies have made.

Plankton and fisheries

The fish of the open seas can be divided roughly into two kinds according to their behaviour. There are those which frequent the sea bed (**demersal** fish), such as

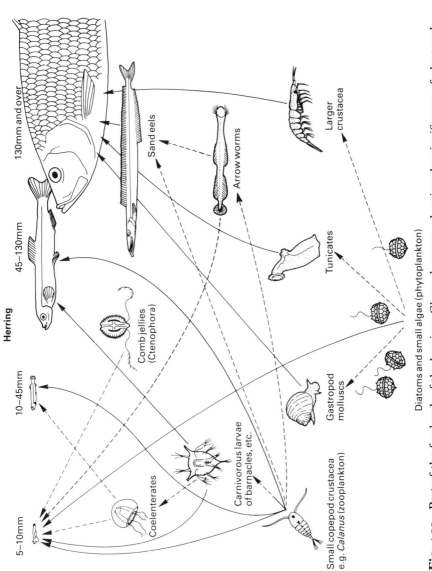

Herring

130mm and over

45–130mm

10–45mm

5–10mm

Sand eels

Arrow worms

Larger crustacea

Comb jellies (Ctenophora)

Tunicates

Gastropod molluscs

Coelenterates

Carnivorous larvae of barnacles, etc.

Small copepod crustacea e.g. *Calanus* (zooplankton)

Diatoms and small algae (phytoplankton)

Fig. 4.11 Part of the food web of the herring, *Clupea harengus*, showing the significance of plant and animal plankton at different stages of its growth. Unbroken lines indicate the food of the herring; dotted lines show the food of other species

cod, haddock, and plaice, whose prey consists almost entirely of bottom-living animals. They are usually caught in trawl nets. However, during their early life, the young fry of these species inhabit the brightly illuminated surface waters where they not only feed on plankton but also form part of it. The second group (**pelagic** fish), which includes herring, mackerel, and pilchard, inhabits the surface layers, either feeding directly on the abundant plankton, or on animals from the next trophic level feeding upon the plankton. Such fish are usually caught in nets suspended from floats at the surface. Thus, in different ways, the distribution and density of both groups are influenced by the distributional pattern of the plankton on which they rely. Such information is of vital concern to our fisheries. The feeding behaviour of the herring provides a typical example and part of its food web is summarised in Fig. 4.11. The youngest fish depend largely on diatoms and the smallest larvae and crustaceans for their food. As their size increases to about 130 millimetres in length, they remain mainly dependent on zooplankton such as the copepod crustacean, *Calanus*, but their diet gradually becomes more varied. Once full grown, they rely less on plankton and seek larger prey, particularly sand eels, crustaceans such as *Nyctiphanes*, and the tunicate, *Oikopleura*. As will be seen from Fig. 4.11, the food of all these species is planktonic.

While a variety of methods such as echo sounding is now used to locate herring shoals, a knowledge of the distribution of plankton can, nonethelss, be a valuable aid in predicting fish distribution and density. In areas where the small crustacean, *Calanus*, is abundant, herring numbers are likely to be high, but when there is an overwhelming mass of plant plankton (known by fishermen as 'weedy water') which is associated with an absence of animal plankton (had herbivores been present they would have kept down the quantity of plants), the catch would

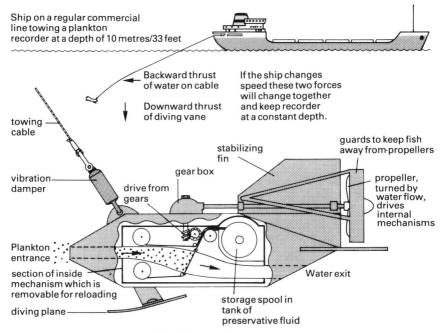

Fig. 4.12   Hardy's continuous plankton recorder

be expected to be poor. More than forty years ago, the biologist Sir Alistair Hardy invented an ingenious device known as the continuous plankton recorder (Fig. 4.12). It consists of a torpedo-shaped cylinder with stabilising fins, and is towed along behind a ship at a predetermined depth and speed. The water enters a small hole at the front of the cylinder, passes through the channel inside and out at the back. On the way, the plankton is filtered out by a moving screen of silk gauze which is wound from one spool to another like a film in a camera. The gauze passes to a third spool in a tank of formalin which acts as a preservative. The mechanism is driven by a propeller located at the rear. Using the calibrations on the gauze, and knowing the location and speed of the ship, and the depth at which the samples are taken, the pattern of plankton distribution in a section of ocean covering several miles can be built up. Over the years, the continuous plankton recorder has proved its worth. At the end of each voyage the instruments are despatched to the Marine Laboratory at Edinburgh, where the spools of silk gauze are unwound, the plankton is identified, and the general picture of distribution is constructed and published. Today, modifications of the original design of the recorder include variable fins so that an undulating course can be covered at different depths, and the provision of electronic devices to record depth, salinity, and temperature.

Elephants and game reserves

In recent years, increasing attention has been paid the world over to the establishment of nature reserves, on account of their tourist and scientific attractions, and also as a means of preserving rare species of animals and plants which might otherwise become extinct through destruction by man or his interference with their natural habitats. However, the preservation of large herbivores such as elephants can pose particular problems. It has been estimated that these animals spend between 17 and 19 hours a day feeding, during which time a single elephant can consume about 150 kg of vegetation. This consists predominantly of grass but also includes leaves of all kinds, some roots, and even the bark of trees when green plant material is scarce. On account of their large size and peculiar feeding habits, their capacity for destruction is almost unlimited, bushes and small trees that stand in their way being frequently knocked down or uprooted. Like all herbivores, they are inefficient utilisers of the energy they consume, the average production of faeces being around 82.5 kg day^{-1} (55 per cent of the food intake). In the areas of Africa and Asia where elephants live, the level of plant productivity is frequently low due to a combination of limited rainfall and poor soil fertility. Elephants feed all the year round and have an average life span of about 65 years. Moreover, as we have seen, their feeding habits are highly adaptable and they are free of predators. In game reserves under the protection of man, their numbers quickly grow and this can impose intolerable grazing pressures on the limited vegetation. In one African national park it has been estimated that herbivorous mammals (mainly elephants) consumed no less than 60 per cent of the total annual net production of plants. This took no account of the physical damage inflicted on the environment, which could well have resulted in a further 10 per cent loss (see Fig. 4.13).

The main problem in running a nature reserve is to ensure that an ecological balance is maintained between the different species present. Excessive competition either within species or between them, will tend to result in the

Fig. 4.13　Grazing pressure by a herd of elephants in an African game reserve (photograph Frank W. Lane)

predominance of some forms and the extinction of others. Quantitative studies of consumption in relation to production, such as have been carried out for the elephant, are thus essential if the aims of reserves are not to be self-defeating. Seen in this context, all populations must be regarded as standing crops (p. 55) whose numbers will fluctuate within certain limits. If the maximum acceptable limits are exceeded, as when the energy demands of a herd of elephants is greater than the production capacity of the environment, it will be necessary to apply checks artificially. Hence the policy of selective culling of older animals which has to be adopted in some game reserves.

Competition

One of the themes running through this chapter and the previous one has been that the resources required by plants and animals are in limited supply. Their availability therefore determines to a considerable extent the density of populations and their distribution. Among plants, competition is particularly intense for light and mineral nutrients, while animals compete for food and shelter. Evidence suggests that different species seldom, if ever, occupy precisely the same ecological niche. Even so, within any locality the availability of a particular kind of niche is bound to be limited, hence competition for it will take place between the individuals of a single species (**intraspecific competition**). Territorial behaviour among animals, particularly birds, as described in Chapter 3, may well have evolved as a mechanism for making the optimum use of the limited food available and hence reducing intraspecific competition to a minimum.

From our previous discussion of the structure of communities, it will be clear that competition also occurs between different species (**interspecific competition**). The interactions involved are often much more complex than those occurring within species, as is illustrated by the structure of food webs. Competition leads to changes in community structure, both qualitative and quantitative, some of which are predictable.

Succession

The principle of interspecific competition and its effects is well illustrated by reference to the colonisation of a bare area of ground such as a recently ploughed field. The primary plant colonists will be mainly annuals, and these are sometimes referred to as an **open pioneer community**. The only animal populations present will be those belonging to soil ecosystems, as well as their bird predators such as starlings and blackbirds. The duration of this phase of colonisation is usually short, perhaps less than two years. The next phase, the arrival of perennials, is more prolonged and may last for around 15 years. The first perennials are herbaceous but later ones are woody. The outcome is a **closed tall herb community**, in which small trees may begin to make an appearance. An abundance of fruits and seeds will encourage colonisation by seed-eating bird species such as finches. The next 15 years or so will see the development of a **scrub vegetation**, consisting predominantly of shrubs and small trees, which has an increasing shading effect on the herb community below. The fruits produced by the larger perennials will be attractive to berry-eating birds such as thrushes, and nut-eaters like jays and pigeons. Meanwhile, the insect population will have increased in numbers and diversified, providing food for insectivorous bird species such as tits and warblers.

The next stage in development, which may occupy anything up to 150 years or more, results in young **broadleaved woodland**, frequently dominated by oak or beech. This may support large carnivores such as owls and buzzards, which prey on smaller birds and mammals. Eventually, an equilibrium will be reached in which the death of old trees is counterbalanced by the maturation of younger ones. This steady state is known as **climax**, the developmental steps by which it is brought about are **seral stages** (each one being a **sere**), and the whole process is referred to as **succession**. The sequence of events described above is shown diagrammatically in Fig. 4.14.

Our knowledge of the successional changes involved in the development of different kinds of communities still remains largely descriptive. Nonetheless, it is possible to summarise in a more objective manner some of the main features common to successions.

(i) The kinds of plants and animals involved in a succession change continuously in a predictable manner, rapidly at first but more slowly later. As we saw in the example of a ploughed field, the first colonists are invariably small herbs (producers), mostly annuals. These must become established before the arrival of herbivorous animals. Carnivores follow, but not until primary production has reached a level sufficient to support them.

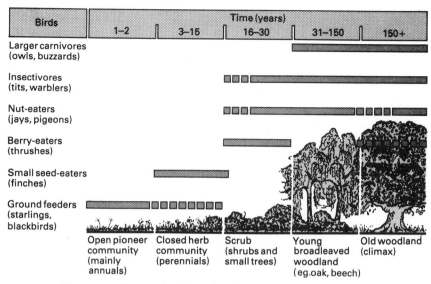

Fig. 4.14 Succession in a ploughed grassland community

(ii) Succession is always associated with increased biomass of living and decomposing plant and animal material. As Odum has pointed out [2.6], an increase in the amount of organic matter and modifications in its composition are two of the main factors controlling the balance of species during a period of ecological change.

(iii) The diversity of species tends to increase with succession. As a generalisation this is sometimes difficult to establish, since the situation tends to vary from one community to another. There is no doubt that it is true of heterotrophs, as is evidenced by the diversity of insects occurring in a mature forest compared with that in young scrubland. The situation among autotrophs is less certain, and it seems likely that the diversity of plant species in many ecosystems may reach its maximum some time before a climax equilibrium has been attained. Nonetheless, in general, ecological niches become increasingly specialised with the advance of successional change.

(iv) Succession involves a progressive decrease in net community production and a corresponding increase in respiration. For growth to occur, total production (P) must increase more rapidly than total energy expenditure (R), i.e. $P > R$. At the theoretical climax, $P = R$, while when degeneration has set in (post-climax) $P < R$. But the number of species present, the biomass of living and dead material, and the P/R ratio continue to remain in a fluid state long after a locality has achieved its maximum primary production.

Time factor in succession

We tend to regard the process of succession as a lengthy affair, a view which is justified in changing land ecosystems such as that described in the previous section. However, many communities exhibit a far simpler structure and their successional climax may be achieved in a matter of weeks rather than years.

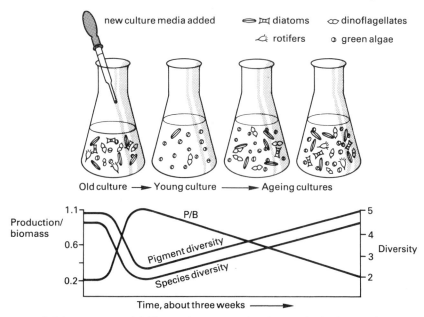

Fig. 4.15 A laboratory model illustrating succession in pond plankton (after Margalef)

Freshwater pools are a typical example, where the primary producers consist predominantly of planktonic algae and diatoms, which provide the food for a limited range of small herbivores such as rotifers and the crustaceans, *Daphnia* and *Cyclops*. Under favourable conditions, the process of succession and the attainment of climax may be achieved several times in a year.

Margalef has devised a convenient way of simulating the interactions of a pond community of planktonic plants and animals under laboratory conditions. The procedure is represented in Fig. 4.15. Starting with an old culture typified by a low level of producers, some new culture medium is added. A young culture is now formed, typified by a high level of production. As succession proceeds, the proportion of animals increases, and productivity gradually declines as climax approaches. The successional sequence is monitored by observing changes in the optical density of the plant pigments present by means of a spectrophotometer. In an old stable culture, the ratio of yellow pigments (430 mμ) to green (665 mμ) tends to be high (between 3 and 5), but the addition of culture solution results in an outburst of small plants (algae and diatoms) so that chlorophyll predominates and the yellow/green ratio drops to around 2. Similarly, in an old culture, the ratio of production (P) to total respiration of both plants and animals (R) tends to approach one, indicating that climax has been achieved. When production is small, its relation to biomass (B) will be low (see the P/B curve in Fig. 4.15). In a young community, production exceeds respiration, so the P/R ratio will be greater than one. Also, the ratio of P/B will be high. As ageing takes place (on the right of Fig. 4.15), the system gradually returns to a state of climax. Under favourable laboratory conditions such changes can be observed and measured in a matter of a few weeks.

Evolutionary implications of succession

Studies of natural selection over the last twenty years or so have established that in circumstances of ecological change the selective elimination of less well adapted individuals within a species can sometimes be far more severe than was previously supposed and may frequently attain a level of 60 per cent or more. We might therefore expect that microevolutionary changes involving variants with survival value could be detectable over a relatively short period of time. This has now been shown to be so, and significant variations in characters with adaptive significance, such as the spot pattern on the hind wings of the meadow brown butterfly, *Maniola jurtina*, correlated with major alterations of habitat, have been detected in a matter of a few generations [4.5].

When studying the occurrence of ecological succession, we tend to overlook the fact that while competition between species (interspecific) must be intense, that taking place within species (intraspecific) can be equally severe. Under such conditions, less well adapted phenotypes will be quickly eliminated, with a consequent adjustment of the gene pool from which the next generation will be derived. During the early stages of succession, a capacity among plant colonists for rapid growth, the maximum exploitation of limited resources, and a high level of reproduction will all be at a selective advantage. These are just the characteristics associated with species commonly regarded as successful weeds. By contrast, survival during the later stages of succession demands such features as large size, a longer life cycle, and an increased ability to cooperate with other species (**mutualism**). Few species manage to survive all the way from the open pioneer community to the climax woodland, but for those, such as the oak, that do, a remarkable degree of adaptability is needed to overcome the changes in environmental conditions. Many of the adaptations will be of a physiological kind, for example modifications in the pattern of growth, which are not easy to study experimentally. However, apart from their biological interest, some could well have economic importance, so this is an area of research with considerable possibilities for further development.

Summary

1 The numbers in plant and animal populations fluctuate, usually within fairly narrow limits. Sometimes fluctuations occur with a regular periodicity; more often they are erratic. Occasionally, catastrophic outbreaks or reductions in density take place.

2 The underlying structure of ecosystems is economic and depends upon a balance between production and consumption.

3 Energy conversion at different trophic levels obeys the First Law of Thermodynamics.

4 Food relationships within a community are complex and vary with the age of the organisms and the seasons.

5 There are limitations to using either numbers or biomass at each trophic level as ways of describing food relationships. Energy transfer from one trophic level to another is the most useful parameter.

6 The fund of nutrients available on the earth is finite. Recycling must therefore occur both of global (atmospheric) and localised (sedimentary) resources.

7 The study of food webs can have great practical and economic value, as in the

fisheries. Conservation, as in game reserves, also depends on a knowledge of food relationships and food availability.

8 Competition within plant communities leads to succession and eventual climax. The process can have profound effects on the animal populations present. Usually the process takes place slowly but it can be rapid, particularly as a result of human interference.

9 Succession can have important evolutionary implications for the populations concerned as it involves powerful selection pressures. Appreciable changes in gene frequency can sometimes be detected in a few generations.

References

4.1 Phillipson, J. (1966) *Ecological Energetics*, Arnold

4.2 Neal, E. G. (1948) *The Badger*, Collins

4.3 Colinvaux, P. (1980) *Why Big Fierce Animals are Rare*, Pelican

4.4 Berrie, A. D. (1972) 'Productivity of the River Thames at Reading', *Symposia of the Zoological Society of London*, **29**, 69–86

4.5 Dowdeswell, W. H. (1981) *The Life of the Meadow Brown*, Heinemann

5 Methods of study

Collecting and sampling

If we wish to study the plants and animals in a population or community, it will usually be necessary to extract a certain number of individuals for more detailed investigation. We might only need to collect a few representative specimens for identification. Alternatively, where quantitative problems, such as the dynamics of a population, are involved, series of samples may be required. Whatever the reasons for our sampling, certain general principles apply.

(i) The number of plants and animals removed should be the minimum required for the purpose. This implies that before beginning any study involving living organisms we should formulate a clear idea of what we wish to find out and how we intend to go about it.

(ii) Repeated visits to the same site, e.g. for project work, can bring about radical changes in the plant and animal communities there, particularly in their density.

(iii) From (ii) above, it follows that, whenever possible, organisms removed from a habitat for study should be returned to it as soon as possible.

(iv) Fieldwork can also have serious effects on the physical nature of a habitat, with drastic consequences for the organisms that live there. Thus, when studying a river, it may be necessary to remove rocks and stones in order to inspect the animals underneath, but these should always be returned, preferably to their former positions.

(v) Finally, sites for study are frequently situated on private land such as farms and woodland. We must be sure to leave them as we find them and particularly to remove any litter.

Collecting apparatus

Nets

Populations of day-flying insects such as butterflies are best sampled by means of a net fitted to a short stick about 600 mm long. Entomological nets are usually of a standard pattern with a jointed frame to permit folding and a diameter of about 400 mm (Fig. 5.1). The net bag is made of terylene, and in order to see the insects inside, it is important that its colour should be black (not green or white as often sold).

For sampling insect communities among low growing vegetation such as on grassland, a **sweep net** is required. Its principal characteristic is robust construction so it will stand up to hard use. Many patterns sold commercially are unsatisfactory for one reason or another, so a successful home-made version is shown in Fig. 5.2(a). The pole is a broom handle approximately one metre long, the net frame is made from a stout metal rod and the bag is of canvas attached to the frame by curtain rings. It is worth remembering that many small animals such as gastropod molluscs and the larvae of butterflies are sensitive to touch and low

Fig. 5.1 Entomological net for collecting day-flying insects

frequency vibrations such as footfall, and react by dropping down into the depths of the vegetation, thus evading capture. Therefore, when sampling, it is important to avoid walking over the area to be sampled (Fig. 5.2(b)).

The designs of nets for collecting in aquatic habitats are now reasonably well standardised, yet it is surprising how many of those sold commercially fail to meet essential demands in some respect or other. By far the most commonly used is the **general collecting net** (Fig. 5.3). Features of its construction are:

(i) it must be robust and able to withstand rough treatment. This applies particularly to the frame holding the net and the mechanism attaching it to the pole.

(ii) the net bag needs to be of a tough material with a mesh size of about one millimetre. This will be small enough to retain all organisms except plankton. Experience has shown that for coarse mesh bags, bolting silk (although more expensive) is preferable to nylon, as it is more durable and also easier to repair in the event of damage.

For collecting plankton a special **plankton net** is needed. This can either be attached to a cord and towed behind a boat or operated by hand on a pole about one metre long (Fig. 5.4). As the net passes through the water if filters off the plankton, much of which is washed down to the base of the cone where it accumulates in the transparent container. Here again, many commercial models are inadequate. The main requirements are:

(i) if the net is attached to a pole, the pole should be of variable length, consisting of several sections that can be screwed together. Collecting can then be carried out at different depths.

(a)

(b)

Fig. 5.2 (a) Design of a home-made sweep net, (b) using the sweep net. It is advisable to avoid trampling on the area of vegetation to be sampled

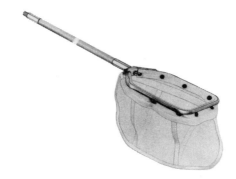

Fig. 5.3 General collecting net for use in water (Philip Harris Ltd)

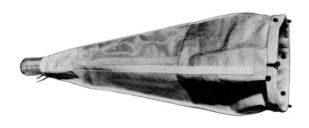

(a)

(b)

Fig. 5.4 Plankton net with (a) a cord attachment for towing, (b) a pole attachment (Philip Harris Ltd)

Fig. 5.5 Drag net for use in aquatic habitats (Philip Harris Ltd)

(ii) the frame must be strong enough to support the weight of water that accumulates in the net when it is full.

(iii) the net must be tough (usually made of nylon) and of appropriate mesh for the purpose. For zooplankton and larger phytoplankton the mesh size needs to be about 0.3 mm. But for small phytoplankton and protozoa an aperture size of about 0.075 mm is required.

(iv) the container should be transparent, unbreakable, easily detached and replaced, and of a standard size so that replacements are readily available.

Bottom living animals that colonise the surface of sand and mud can be sampled by means of a **drag net** (Fig. 5.5) which can also be used for grassland communities of animals provided the ground is reasonably flat. The equipment consists of a strong metal frame to which is attached a tough bag about 500 mm deep. Coarse mesh nylon is the best material for collecting in water, while canvas or linen are equally good for use on land. The net is pulled along by a cord attached to the frame, which is so designed that it runs as smoothly as possible over

Fig. 5.6 Using a home-made drag net to sample a grassland community

the substratum. As with the sweep net, when collecting on land care should be taken to avoid walking on the area to be collected, otherwise a non-random sample will result (Fig. 5.6).

Beating apparatus

Interesting studies can be made of the animals inhabiting small bushes and trees in order to determine the extent to which they represent distinct ecosystems, for instance by finding the ratio of herbivores to carnivores. This necessitates dislodging the organisms by beating with a stick and collecting them as they fall. An old sheet or large piece of white plastic can serve as a receptacle but these are sometimes difficult to manipulate, particularly in a wind. The standard equipment designed for the purpose is the **Bignell beating tray** (Fig. 5.7). This consists of a collapsible wooden frame supporting a piece of strong fabric about one metre square. A satisfactory alternative for many purposes is an inverted umbrella.

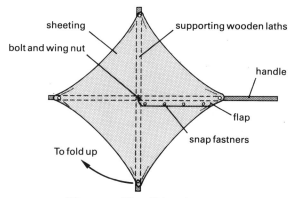

Fig. 5.7   Bignell beating tray

Pooter (aspirator)

Sampling communities of small insects inhabiting rough surfaces such as the bark of trees poses particular problems and there is no easy way of achieving randomness. However, provided an element of selection is permissible, the pooter (Fig. 5.8) is a useful device, and can easily be constructed from a specimen tube. Suction is usually by mouth, but a rubber bulb is much more convenient.

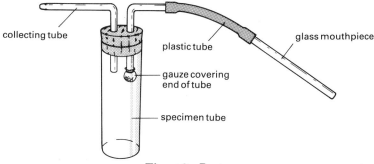

Fig. 5.8   Pooter

Traps

(a) Pitfall traps

At their simplest these consist of containers such as tins or jam jars sunk below the surface of the ground so as to create the minimum disturbance to the environment (Fig. 5.9). They are then covered over thinly with sticks and other vegetation appropriate to the area. Since they are likely to be invisible, it is important to mark their whereabouts. They are particularly effective for catching the larger arthropods such as beetles, centipedes, millipedes, and spiders, most of which are active at night, but they do not provide reliable estimates of density. A number of pitfall traps set up in the same area can give a good idea of the local community. Moreover, marking the animals with dots of cellulose paint and releasing them after capture can provide evidence of the extent of movement. For maximum effectiveness, a battery of traps should be left in position for several days, so that the animals become accustomed to any disturbance of the terrain that may have occurred in setting them up.

(b) Mammal traps

Most small mammals are nocturnal, so the usual practice when sampling populations is for a night's capture to be removed the following morning. However, some species such as voles are active both by day and night, so on occasions, a stay in the trap of several hours may be unavoidable. If appropriate precautions are not taken a high mortality may result during cold weather. The problem can be overcome to a considerable extent by providing a nest box containing dry hay or bracken and a supply of mixed grain. Various baits can be used, one of the most attractive being coconut, which should be renewed at frequent intervals. By far the most successful design of trap is the Longworth mammal trap. This is constructed of light alloy (Fig. 5.10). The nest box is detachable from the body of the trap, which greatly facilitates cleaning and the removal of captive animals. The animals are best transferred to a cloth bag (not plastic which can be airtight) of suitable size before inspection. Other features of the Longworth trap are:

(i) a semi-automatic method of setting which is independent of individual differences in skill.

jam jar sunk below ground level

covering of sticks and leaves

Fig. 5.9   Pitfall trap

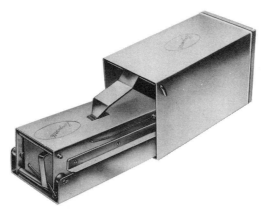

Fig. 5.10 Longworth mammal trap (Griffin and George Ltd)

(ii) adjustable sensitivity to weights (trip tension) ranging from 1.5 to 8 g.
(iii) positive closing of the trap door which moves downwards and forwards from within the trap, not backwards from outside it.
(iv) a pre-bait safety catch which enables the trap to be left open to permit preliminary exploration by animals prior to setting.
 (v) alloy construction results in light weight, considerable durability, and ease of transport and cleaning.

(c) Water traps
These are plastic or metal bowls which are filled with water, to which a little detergent and preservative (e.g. formalin) have been added. They can be coloured if desired. The detergent is important as it reduces the surface tension and insects alighting on the water thus tend to sink. The traps can be set in a variety of places and are particularly useful for studying insects such as water crickets and mayflies that settle on the surface of ponds. For this purpose the traps are best arranged floating in the water with suitable anchors.

(d) Light traps
Traps using short wave light are now the commonest means of capturing insects particularly moths, at night. A variety of designs are on the market, that shown in Fig. 5.11 being particularly convenient on account of its portability. The mercury vapour lamp operates off a 12 volt car battery. The whole unit is collapsible and lightweight, and designed to be carried, together with its power supply, in a rucksack. In deference to the English climate, it is fitted with a drain funnel in case of rain. The insects are attracted to the light and become stunned on impact with one of the baffles, falling down the funnel into the collecting box below. Subsequent handling of the catch can be greatly simplified if portions of egg cartons are stacked round the base of the funnel to which the insects will tend to cling. Some light traps contain a killing agent, but, as stressed at the beginning of this chapter, whenever possible the insects should be kept alive and released again after inspection.

Fig. 5.11   Portable light trap for catching insects at night (Griffin and George Ltd)

It must be remembered that light traps cause an upset in the distribution of the insect populations within their range. Like the pitfall traps mentioned earlier, while they provide some idea of the different species present in an area, they are not a reliable means of estimating population densities. It is also worth remembering that light of certain wavelengths kills some species of insects.

Widger

When handling samples of soil or small animals there is often a need for an appropriately shaped spatula. For the smaller organisms there is little to beat a piece of stout copper wire, heated at one end and hammered out to form a flat surface. For larger specimens and solid materials, the widger, much beloved of glasshouse gardeners for pricking out seedlings, is ideal (Fig. 5.12). It is a piece of stainless steel 180 mm long, with two concave blades of different widths (roughly 6 mm and 20 mm respectively).

Fig. 5.12   A widger (Griffin and George Ltd)

Extraction of soil organisms

The occupants of soil and the litter layer above it often exhibit characteristic distributions, but since we cannot see them, we are frequently unaware of the fact. Moreover, populations vary greatly in density depending on such factors as humidity, pH, and the organic content of the soil.

1 Sampling earthworm populations

Darwin estimated an average earthworm population, mainly *Lumbricus terrestris*, at about 12 000 in 1000 square metres, but we now know that densities ten times this number are quite common. Moreover, it is clear that variations in density by a factor of 100 or more can occur even within a few yards. Physical factors and soil textures are partly responsible but we must also remember that large local aggregations of young worms may build up during the breeding season. How long these take to disperse is not known.

All worms are burrowers but the depths to which they penetrate vary greatly both with the species and the soil conditions. While forms such as *Lumbricus rubellus* are shallow workers, seldom penetrating more than 10 cm below the surface, larger species like *Allolobophora nocturna* and *Lumbricus terrestris* are found at depths of 100 cm or more. The distribution of different species and their varying behaviour have been described elsewhere [5.1]. Typical data covering a few examples are shown in Fig. 5.13.

The extraction of earthworms from soil presents considerable difficulties and no technique has yet been devised which achieves total success. Various chemicals, when applied in solution, evidently act as irritants and cause the worms to come to the surface. One of the most effective is methanal (formaldehyde) which should be used as a 2 per cent solution (25 cm³ of 40 per cent methanal in a garden watering can filled to capacity (usually about one gallon) – the exact concentration is not

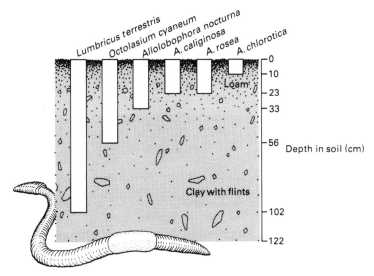

Fig. 5.13 Vertical distribution of some species of earthworms in a Rothamsted pasture (after Leadley Brown)

critical). The can should be fitted with a rose and its contents applied to a one square metre area of ground, which can be marked out with string. A limitation of this method is that penetration of the soil by methanal seldom exceeds one metre, and even this depth is achieved only slowly. Some of the deeper burrowing species such as *Lumbricus terrestris* tend, therefore, to be overlooked. Where possible, methanal sampling should always be supplemented by:

(i) digging of vertical profiles to check the distribution of the different species.
(ii) the setting up of a wormery (a wooden box with glass sides) in which different species can be kept under conditions resembling as closely as possible those in the field. Their behaviour and mode of burrow formation can then be observed and checked against those found in the field.

Although earthworm densities are still often quoted as numbers per hectare, such figures mean relatively little when applied to small areas of a square metre or so where large fluctuations can occur over a short distance. Furthermore, although no method is known of achieving total extraction, this does not necessarily invalidate the use of the methods outlined above, which are valuable for comparative purposes provided they are used in a consistent and standardised manner.

2 Extracting small organisms

(a) Flotation

Small arthropods, greater than about one millimetre in length, can be removed visually from soil and litter by floating them on liquids whose specific gravity is slightly greater than that of water, such as 25 per cent salt solution (specific gravity 1.19). The success of this procedure depends largely on taking a small amount of soil at a time, stirring it thoroughly with the flotation fluid and allowing it to settle for a few minutes. Many larvae and other small arthropods are dark coloured and therefore show up against a white background. Others, such as springtails, are white, and are best seen against dark surroundings. The ideal container is therefore translucent (glass or plastic) so that it can be viewed against a variety of backgrounds. In practice the use of salt solution has unfortunate osmotic effects on some species such as springtails, causing them to disintegrate quickly, and for approximate comparisons much can be achieved by way of extraction using ordinary tap water in which the animals float on account of the air in their tracheae. Like all techniques, this one has to be learnt and practised; at first people frequently assert that they can see nothing but liquid mud! Before making comparative estimates, it is important to 'get your eye in'.

(b) Tullgren funnel

Soil organisms inhabit an environment that is predominantly damp, dark, and cool; they tend to shun conditions that are the reverse of these. This is the principle behind the Tullgren funnel (Fig. 5.14) in which a sample of soil is illuminated and warmed by a low-power electric light bulb, the animals being collected in a dish of preservative underneath. Various commercial versions are available, and a typical example is shown in Fig. 5.15. Although it does not achieve 100 per cent extraction, the Tullgren funnel is, nonetheless, an extremely efficient

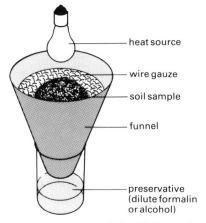

heat source

wire gauze

soil sample

funnel

preservative
(dilute formalin
or alcohol)

Fig. 5.14 Principle of the Tullgren funnel

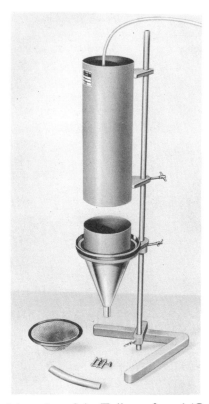

Fig. 5.15 Commercial version of the Tullgren funnel (Griffin and George Ltd)

apparatus if used properly – which all too often it is not. Some common causes of failure are:

(i) too high powered electric light bulb. The maximum should be 25 watt and usually 10–15 watt will suffice.
(ii) electric bulb too near to the soil sample. This overheats the animals and kills them before they can escape downwards.
(iii) too thick a sample of soil. This means that much of the soil will have been dried and the animals killed before they can escape. Whenever possible, use a number of Tullgrens each with a small soil sample rather than one with a large one. Incidentally, this arrangement provides a useful means of checking the numbers extracted in different samples from the same place.
(iv) the angle of the cone below the soil sample is too broad and the aperture too narrow. Both have the effect of causing animals to adhere to the surface, thus failing to reach the collecting vessel. A satisfactory laboratory model can be made using a paper cone held in position by a tripod, with a circular piece of wire gauze to support the soil or litter sample.

(c) Baermann funnel

In addition to the myriads of protozoans which inhabit the water in the spaces between soil particles and in the film on their surface, a number of other small animals, such as nematodes, rotifers, and enchytraeids, are closely associated with soil water. These cannot be extracted by a Tullgren funnel since their powers of movement are limited and the slightest reduction in humidity soon kills them. The Baermann funnel (Fig. 5.16) makes use of the same principle as the Tullgren but the soil or litter sample is enclosed in a coarse mesh bag and suspended in a funnel filled with water. The organisms move downwards and collect in the vicinity of the

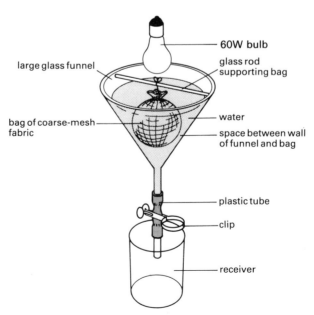

Fig. 5.16 Principle of the Baerman funnel

spring clip whence they can be released into the collecting jar below. Since there is no danger of desiccation and an appreciable volume of water to be heated, a more powerful electric light bulb can be used – usually about 60 watt.

Comparing the different methods of extraction available for use in the laboratory, Leadley Brown [5.1] concludes that funnel extractions are better suited to organic soils and litter, while flotation methods, although more time-consuming, are more efficient for mineral soils. For a more detailed analysis of many methods of study used in quantitative soil ecology, reference should be made to Phillipson (1971) [5.2].

Sampling methods

The problems of investigation and experimentation in biology have been set out and discussed by Heath [5.3], and for a concise overview of the subject as a whole the student would do well to consult his account. Here we are concerned with a particular aspect of investigation which constantly faces ecologists. Populations and communities, as we have seen, are usually 'large', and this nearly always precludes the possibility of observing every individual present. We therefore need to obtain one or more samples, and in doing this we need to ensure as far as possible that those individuals extracted are representative of the group to which they belong. Statistical methods (described in Chapter 7) enable us to compare samples and to make estimates of the features (**parameters**) which characterise populations. Most features of species, such as size, shape, and colour, vary; hence they are known as **variates**. Mathematical procedures exist which enable us to gauge and compare the overall range of a variate, one of the most useful being the arithmetic **mean** (μ) or numerical average. We also need a parameter that specifies the degree of variability of values (**dispersion**) and here we have the **standard deviation** (σ) and its square, the **variance** (σ^2). While the ecologist need not be concerned with the mathematical origins and derivation of the various formulae associated with these statistics, it is nonetheless essential that he or she should understand the circumstances in which particular statistical procedures are appropriate, and still more, the occasions when they should not be used. These issues are discussed in Chapter 7.

To be effective, all sampling methods must conform to the following essential requirements:

(i) they must be repeatable (i.e. the technique can be standardised).
(ii) in order to ensure repeatability, they should be as simple as possible so as to reduce to a minimum errors introduced through the application of the sampling technique.
(iii) the mode of sampling and nature of samples (e.g. their frequency and size) must be closely related to the method of analysis to be used. When planning an investigation, sampling procedure and the subsequent handling of the data obtained should be considered together.

1 Crude procedures

It is impossible to specify an appropriate sampling procedure for every kind of plant and animal habitat because they vary so greatly. However, certain guiding principles apply in most circumstances. At the beginning of an ecological study of a population or community, it is helpful first to gain an approximate idea of the general situation before adopting more detailed sampling procedures. Thus a quick survey of the diversity and density of mayfly nymphs in a stream could be obtained from counts of individuals underneath stones of an 'average size'. Again, densities could be categorised roughly into high, medium, and low.

Sampling populations of small animals, such as the freshwater shrimp, *Gammarus pulex*, inhabiting aquatic vegetation and mud can pose particular problems. Here a standardised kick or shuffle can be used to dislodge the animals which are then collected in a net placed downstream. We should not be afraid of devising such crude sampling methods of our own, provided that they conform as far as possible to the requirements outlined above, and that we are aware of their likely shortcomings. One of the advantages of statistical procedures is that they enable us to assess the magnitude of our sampling errors and hence to compare these with differences observed between the variants we are studying. If these differences prove to be less than those of the sampling methods used in studying them, the sampling procedures are clearly not sensitive enough for their purpose.

2 Quadrats

One way of standardising sampling procedure which is particularly appropriate to the study of plant communities is the **quadrat**. A quadrat is an area of known size. It is usually a square but can be circular or any other shape required. Its size will depend on the type of vegetation to be sampled and the kind of information needed. Thus, for a survey of an oak wood, involving a variety of species, it will be desirable to use fairly large quadrats of, say, a 20 metre side. These can be marked out with string which is pegged down at the corners. For ease of counting it is customary to subdivide quadrats into smaller units. Thus a one metre quadrat is divided into 100 mm squares. In order to take account of seasonal effects, for instance the disappearance of annuals and herbaceous perennials during winter, and also for long term studies of change, it is desirable to establish **permanent quadrats** which can be visited at prescribed intervals. For studies of the density of plant populations a one metre quadrat is usually used. This consists of a light wooden frame with cross wires marking the subdivisions. Transportation is difficult and the most convenient version is hinged at the corners so that it will fold, the subdivisions being marked by elastic which stretches taut when the quadrat is open. Again, for sampling populations of mosses and lichens on a wall or small animals inhabiting the surface of stones in a stream or the rocks on a shore, a 200 mm quadrat may well suffice, subdivided into 50 mm squares.

The use of a quadrat as a sampling device necessitates certain decisions:

(i) the size of quadrat needed. This is related to the size of the unit to be studied. Thus, if we are sampling a wood, a small quadrat will include only a few trees in spite of the fact that these are the dominant species exerting the greatest ecological effects. If a quadrat is too small, human errors due to the 'edge effect' will be greatly increased – that is to say the problem of deciding whether plants on the edge should be included or not. It should also be remembered that the larger the quadrat, the greater the labour of counting the individuals inside it; an important consideration if time is limited.

(ii) the number of quadrats required. The fewer the samples we take, the less our results will be representative of the population as a whole and the greater will be our sampling error. In studies of this sort we are usually attempting to fit the samples to a curve of normal distribution (see p. 131). As will be explained in Chapter 7, if the number of observations is small, values for the **standard deviation** and **standard error** will be correspondingly large. We may therefore be faced with a situation where the population mean deviates so widely from that of our sample that it is impossible to make any valid deductions regarding the nature of the population as a whole. Statistical procedures thus provide a useful guide to the sensitivity of sampling methods and can frequently help in assessing the question of whether further quadrat samples are necessary or not. For this reason, an essential piece of equipment for all students of ecology is a pocket calculator so that these elementary calculations can be performed *on the spot*. Once back in the laboratory it is probably too late. For studies of population density, it must be remembered that the bigger the number of individuals counted, the greater will be the accuracy. It therefore follows that the lower the density being studied, the greater the number of quadrats required.

(iii) the placing of quadrats. In circumstances when an ecological environment is subject to alteration (frequently through human interference) such as the clearing of low growing vegetation in a wood or the planting of conifers, radical changes can occur in the adjacent plant and animal populations. The monitoring of such changes over a period of several years can be of great interest in illustrating the early stages of succession and the speed with which they occur. It can also provide a useful opportunity of comparing changes predicted on some hypothesis with what actually happens. For this kind of exercise one or more permanent quadrats should be laid out, their siting being determined by the objectives of the study.

No population, however, is distributed uniformly, so if we wish to estimate such parameters as density and dispersion using quadrats, we must adopt some system of random sampling. This is necessary because randomisation is one of the assumptions made in most forms of statistical analysis such as those described in Chapter 7. It might be thought that a sure way of avoiding human bias would be to shut one's eyes, throw a quadrat in the air and collect the information from the place where it lands. This procedure certainly helps to reduce the human factor, yet mathematical analysis has shown that the results so obtained often depart appreciably from random.

One of the easiest ways of achieving randomness when sampling vegetation is to set up two lines at right angles to one another so that they straddle the area of study. The layout of quadrats is determined and the position of each is located by measurement (pacing will be accurate enough) along the *x* and *y* axes. The values of each pair of coordinates can be obtained from a table of random numbers which will be found in most statistics textbooks [5.4, 5.5].

A disadvantage of this sampling procedure is that while it achieves randomness, in a diverse area, it often results in quite important regions remaining unsampled. The resulting samples are therefore far from representative of the locality as a whole. For class or project work, when time is always a limiting factor, this may be the most that can be achieved. For really meaningful results a more systematic approach is necessary. Ashby [5.6] has illustrated one such scheme by reference to the zonation of a single plant species in the vicinity of a rabbit warren (Fig. 5.17). An arbitrary base line is established which borders the locality to be sampled.

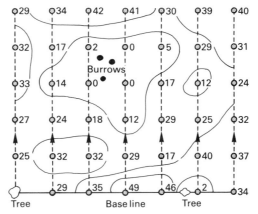

Fig. 5.17 Hypothetical results of sampling a plant population in the vicinity of a rabbit warren. Forty-one quadrats are laid down on a grid, the site of each being represented by a small circle (). The figures represent percentage frequency recorded in each quadrat (after Ashby)

From it, perpendicular lines are run off at regular intervals, using human paces to measure distance and a compass to judge the 90° angles. The quadrats are then sited at regular intervals along the perpendicular lines. This procedure tends to be time-consuming, but it has the advantage of

(i) overcoming the human factor in the choice of sampling sites,
(ii) a more systematic and complete coverage of the area of study.

A useful parameter for measuring the success achieved by different plant species in colonising the same habitat or the establishment of the same species in different places, is **percentage cover**. As we have seen, one of the difficulties in using quadrats to gauge this and similar measurements is the edge effect due to the quadrat only partially covering individual plants or aggregations of them. Clearly, the smaller the quadrat, the less this effect will be evident, until at pin-point size it will virtually disappear. Every 'score' then becomes a hit or a miss. Thus, if we let down a series of pin-points onto the vegetation at random and record the number of hits and misses for a particular species, provided the sample is large, the percentage cover will be:

$$\text{Percentage cover} = \frac{\text{Hits} \times 100}{\text{Hits} + \text{Misses}}$$

The most convenient apparatus for this kind of sampling is the **point frame** (Fig. 5.18). This consists of a wooden framework holding a series of sliding pins (it is important that the pins should slide freely). Some typical dimensions are shown in the diagram. As will be seen, the apparatus can be used both on smooth and rough ground provided any discontinuities present are not too extreme. To ensure random sampling, the same considerations apply as for quadrats. Clumping of plants, such as a patch of clover on a lawn, may present difficulties in scoring and here it is probably best to record for one pin only (which pin to use can be decided at random) for each new sampling position of the frame.

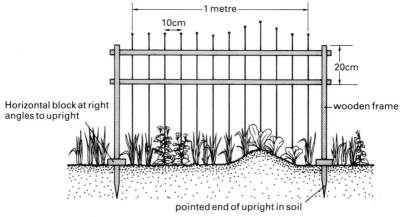

Fig. 5.18 Design and use of a point frame

Much has been written on the mathematics and quantitative techniques of sampling by quadrats, and the above account is no more than an introduction. For further information the reader should consult one of the more advanced works on the subject [5.7].

3 Transects

If we wish to study the changes in plant populations that take place, say, in a succession from water to land, or to compare two contrasting localities, we can do so by setting up a series of quadrats end to end. We then have a **belt transect**. An advantage of belt transects is that they provide a means of recording with considerable accuracy the species occupying a limited zone, their frequency, and cover. However, if employed for more than a short distance, counting can be exceedingly laborious and time-consuming. In practice they are often set up with two parallel pieces of string, say a metre apart. For ease of scoring it is customary to divide the belt into one metre squares (quadrats).

A variation on this theme which is a good deal less sensitive as a method of sampling is the **line transect**. As its name implies, this consists of a line laid along an area of ground, scoring being restricted to those species actually covered by the transect. The result obtained may, therefore, be nowhere near representative of the community as a whole. However, the method has the merit of being quick and can provide a non-quantitative idea of some of the more typical species present. It is particularly useful in illustrating the main features of a zonation such as that occurring on a rocky shore or a succession from a water community to a land one (**hydrarch succession**). The results of transects can be presented either in the form of a table or as a profile chart (Fig. 5.19).

An important consideration concerning all sampling methods, transects in particular, is that before selecting any one of them, the reasons for making the choice must be clearly thought out. These include the objectives of the investigation, the kinds of information required, and the intended method of handling the data obtained. All too often the results obtained from transects are purely subjective and amount to little more than unqualified lists of species.

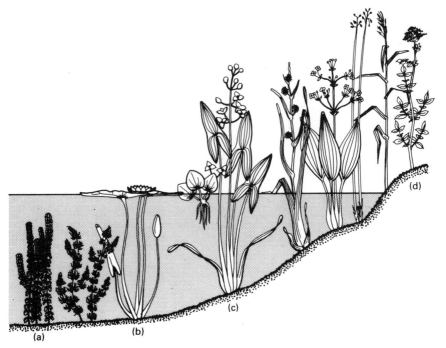

Fig. 5.19 Profile chart derived from a line transect of a pond showing the sequence of changes in a hydrarch succession: (a) submerged plants such as Canadian pondweed (*Elodea*), (b) aquatic species with floating leaves, e.g. water lily (*Nymphaea*), (c) species rooted in water with aerial parts, e.g. arrowhead (*Sagittaria*), (d) true land plants with a preference for humid conditions, e.g. meadowsweet (*Filipendula*) (after Ashby)

Marking animals

Unlike plant populations which are static, many animal species are highly active, and their mobility presents particular problems in the study of their distribution and density. These problems can to some extent be overcome when determining patterns of movement or population numbers by the use of appropriate marking methods.

The traditional way of marking birds is by the attachment of rings to their legs. However, trapping and ringing are now strictly controlled and can no longer be used by anyone. Similarly, the ringing of small mammals and the tagging of fish are specialist techniques, as is the use of radioactive isotopes [5.8].

By far the most effective way of marking many animals for a short period of time is by means of paint or dyes. Quick-drying cellulose paint is available in a variety of colours and its solvents are non-toxic if used in small amounts. The paint is easily applied with any pointed object, such as the head of a small pin inserted into a wooden handle or a sharpened matchstick. For small insects such as the flour beetle, *Tribolium*, a human hair has proved a satisfactory means of application. It is worth remembering that paints do not adhere well to shiny surfaces. Thus, when marking the shells of snails (always on the lower side to avoid attracting the attention of birds), a small patch of periostracum must first be removed. This can easily be done with a nail file or the blade of a sharp knife. An invaluable home-made electrical device for use in the field is shown in Fig. 5.20. This consists of a

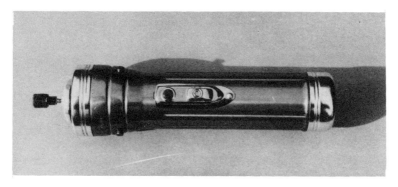

Fig. 5.20 Simple device for removing the periostracum from snail shells. The bulb unit of the pocket torch is replaced by a small electric motor with a carborundum head attached to the spindle

two-cell pocket torch in which the bulb unit has been replaced by a small toy electric motor with a carborundum head soldered onto the spindle. In practice, two cells have proved sufficient for the marking of around 1000 snail shells.

A wide variety of dyes and fluorescent substances have also been used as markers. Non-fluorescent compounds such as aniline dyes can be applied to the bodies of insects as dusts or in solution. Fluorescent substances like zinc sulphide powder are easily detectable under ultra violet light, a suitable adhesive additive being gum arabic. Alternatively, insects such as bees can be dusted with fluorescent powder mixed with a carrier such as talc.

One of the simplest marking devices now available is the felt-tipped pen, and this will be found ideal for insects such as butterflies. Bearing in mind the effects of the weather, it is desirable to avoid water-soluble ones.

Estimating animal populations

The sampling methods described so far are appropriate for plants and relatively static animals. However, since many animal species are active, estimating the density of their populations requires special techniques.

1 Direct counts of large animals

An estimation of population size by direct counting depends on the assumption that a high proportion of the individuals present are visible at the same time. This will only be true of relatively large animals, so the procedure has been used mainly in the study of birds, and mammals such as seals and rabbits. The best time of the year to count a bird population is during the breeding season when movement is restricted by territorial behaviour (p. 37). The accuracy of counting can be greatly increased by employing a number of observers walking in line abreast, each being responsible for the particular sector directly in front. It is important to record all sitings *directly they are made* and to make allowances for animals entering and leaving the area. The procedure is used extensively in the study of populations of game birds such as partridges but it can be used effectively for any species inhabiting relatively open country.

In woodland and other habitats with an abundance of cover, the problem of counting becomes more difficult. A number of new variables, such as the rate of movement of the observer and of the birds, and the range at which the birds can be detected, are introduced. Yapp has proposed a 'theory of line transects' [5.9] and this deserves more consideration than it has so far received. The problem of encounters between moving objects, such as an observer and a bird, is essentially the same whatever the nature of the objects concerned. The frequency of an encounter depends upon:

(i) the size of the objects
(ii) their speed
(iii) their density.

If an observer moves in a particular direction counting all the animals he sees of a single species, then, making a few approximations, the number of individuals per unit area (D) can be calculated from the formula:

$$D = \frac{z}{2R(\bar{u}^2 + \bar{w}^2)^{1/2}}$$

where z = the number of encounters between observer and organism in unit time
R = the range at which the organism can be recognised
$\bar{u}$ = the average speed of the organism
$\bar{w}$ = the average speed of the observer.

When applying this technique, certain considerations must be taken into account:

(i) as with all estimates of density, numbers must remain reasonably constant during the period of study. Resident populations of birds tend to be at their most stable during the summer and this is the best time to determine their density.
(ii) the speed of the observer ($\bar{w}$) can be fixed arbitrarily at about 3 km h^{-1}. Pauses to make records must be taken into account but these should be reduced to a minimum.
(iii) the speed of the organism ($\bar{u}$) varies greatly from one species to another. For instance, the average for tits is between 0.2 and 3 km h^{-1} (say 1.5 km h^{-1}).
(iv) the range at which the bird can be recognised (R) can be either visual or auditory, the latter often being the easier to measure, particularly with birds. For many species this is about 200 metres. Thus, moving at 3 km h^{-1} (i.e. covering approximately 200 metres in three minutes) observers should have a reasonable chance of encountering the majority of woodland species in a single transect.
(v) the number of encounters in unit time (z) is a function of the other variables already considered. In any study, the denstiy of the population over a short period (D) and the average speed of the observer ($\bar{w}$) will remain reasonably constant. The range of recognition (R) and the average speed of the organism ($\bar{u}$) are likely to vary considerably from one situation to another.

For valid estimates of density and comparisons of population size it is worth remembering that:

(i) weather conditions, the time of day and season when the counts are made must be reasonably similar

(ii) features of the environment affecting recognition such as the presence or absence of leaves must be uniform.

The line transect method for estimating densities opens up numerous possibilities for experiment with different species in varying circumstances. It is a method which can be used in the study of animals other than birds, such as squirrels. When conditions are carefully chosen so as to provide consistent information on the variables involved, there is no reason why it should not provide a satisfactory means of making quantitative comparisons of density both between different species in the same habitat and a single species in differing environments.

2 Density of small organisms

As we have seen (p. 62), populations of planktonic organisms, both plant and animal, occupy an important position in many food chains, so estimates of their density in varying circumstances can be an essential part of any study of population dynamics. Sampling can be carried out with a plankton net under controlled conditions (p. 73) and numbers in the catch determined by counting in a cavity chamber (Fig. 5.21) containing 100 mm^3 of liquid. This can easily be made in the laboratory but commercial versions are also available. A microscope slide provides a good base and should be ruled into ten 10 mm squares, the lines being made with a diamond and then rubbed over with a black wax pencil to make them more visible. The sides of the chamber are made from strips of microscope slide glass cemented onto the base with a resin glue such as 'Araldite'. A small volume of the water sample is introduced into the chamber on the slide and a long cover glass lowered gently on top, care being taken that no air bubbles are trapped inside.

The organisms can be counted under a low power binocular microscope. Sometimes it will be a question of counting individuals of a single species; at other times information on the relative abundance of different kinds may be needed. When making estimates of the second kind it is important not to try to count too many species at once; five is generally the maximum. The procedure for counting is to record the number of each species in each 10 mm square until a total of, say, a hundred observations have been made. The occurrence of different species is then expressed as a **percentage frequency**, i.e. as a percentage of the total number of observations made.

When applying this technique, the following points should be borne in mind:

(i) in order to ensure randomisation of the catch, always shake the sample before transferring a portion to the counting chamber with a bulb pipette.

(ii) it will frequently happen that the density of the sample is too great for accurate counting. Appropriate dilution will, therefore, be required.

(iii) since the distribution of plankton in water is never uniform, when estimating the density of populations in a particular locality it is important to obtain counts from as many different samples as possible.

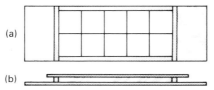

Fig. 5.21 Cavity chamber: (a) surface view, (b) side view

3 Estimation of numbers by mark, release, and recapture

Highly mobile species of animals such as flying insects and fishes pose particular problems where the estimation of their population density is concerned, and neither of the procedures described above is appropriate for the purpose. The technique of mark, release, and recapture has proved useful in such circumstances and is essentially an extension of the sampling methods outlined earlier.

Suppose we are studying a colony of animals such as butterflies and we catch 50 individuals and mark each with a dot of paint or by some other means (see p. 90). They are then released into the population and allowed to randomise for, say, 24 hours. A further sample of 40 is then taken and found to contain 10 specimens marked the previous day. The *flying population* can then be estimated as:

$$\frac{50 \times 40}{10} = 200$$

This is known as the **Lincoln Index** and can be written as

$$P = \frac{an}{r}$$

where P = population estimate
a = number of marked individuals released
n = number in the subsequent sample
r = number of marked individuals in the subsequent sample.

With small samples of less than 20, it is advisable to make a correction to the formula (Bailey's correction):

$$P = \frac{a(n + 1)}{r + 1}$$

The density of populations never remains constant for long. However, over short periods of a few days or so it is not likely to fluctuate greatly and in the interest of obtaining a more accurate estimate of numbers it may be desirable to obtain samples on several occasions. **Bailey's Triple Catch** method is based on this supposition. A population is sampled on occasions 1, 2, and 3, the catches being n_1, n_2, and n_3 individuals. Of n_1 and n_2, a_1 and a_2 respectively are marked and released. (Usually $a_1 = n_1$ and $a_2 = n_2$ unless any individuals die or are damaged in marking and have to be destroyed.) The data can be summarised most conveniently in the form of a trellis (Fig. 5.22).

Continuing our sampling, of n_2 (sampled on occasion 2) r_1 were already marked and r_2 unmarked, and of n_3 (sampled on occasion 3) r_3 were unmarked, while r_4

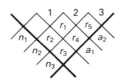

Fig. 5.22 Trellis method of recording data from Bailey's Triple Catch method of estimating animal numbers

were marked on occasion 1 and r_5 marked on occasion 2. The size of the population can then be calculated as:

$$P = \frac{a_2 n_2 r_4}{r_1 r_5}$$

As in the Lincoln Index described earlier, an adjustment must be made for small samples of less than 20 and the corrected formula is then:

$$P = \frac{a_2(n_2 + 1)r_4}{(r_1 + 1)(r_5 + 1)}$$

If required, capture-recapture methods can enable us to follow the fortunes of a population over an indefinite period; the principles involved are the same as those already described but the method of recording becomes a little more complex.

Suppose we wish to estimate the fluctuations in a population of butterflies over a period of about a fortnight – the maximum expectation of life in some species. The records of sampling and recapture are most conveniently summarised in the form of a trellis diagram (similar to the Bailey Triple Catch method) with the dates of the samples running along the top. From each date, lines run downwards south-east and south-west at 45 degrees so as to intersect, as illustrated in Fig. 5.23.

The daily samples are entered at the end of the column running south-west from the data in question, while the totals released appear in the corresponding position to the south-east. The two numbers will normally be the same except in the event of damage and deaths. All recoveries are shown in the body of the table and the number of marks they carry, together with those of insects marked for the first time, represent the total marks released on each occasion. With this procedure it is not necessary to collect every day; in the table a dash shows that no recaptures were possible, either because no insects were caught or because none were released the day before.

Simple direct estimates can be made from the table of the daily flying population. For example, consider the line running south-west from 17th August: out of 9, 6, 9, 4, 9, 4, 2, 5 (total 48) butterflies caught on subsequent dates, 4, 2, 1, 1, 5, 1, 1 (total 15) recaptures belonged to the 13 marked on 17th August. Thus the total number flying on that date was approximately (48 × 13)/15 = 41 insects.

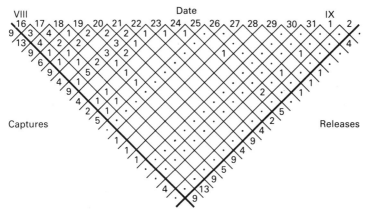

Fig. 5.23 Results tabulated as a trellis in the capture-recapture method of estimating animal numbers over a prolonged period

For the effective use of all mark, release, and recapture methods certain requirements must be fulfilled:

(i) the species used must be one in which marked individuals will randomise regularly when released.
(ii) the sampling method used must be random.
(iii) the method of marking needs to be of a kind which can distinguish between the different dates of capture such as dots of paint in varying colours and positions.
(iv) the populations to be studied must be reasonably stable and not subject to rapid fluctuations in density. But as explained earlier, no population remains static for long. Immigration and the appearance of young will tend to increase numbers while emigration and deaths have the opposite effect. In studies of any duration, therefore, it is necessary to compare the number of recaptures *expected* in a population distributed at random with those actually *obtained*, using as a basis the number of *marks* existing from day to day. The analysis of such data is beyond the scope of this book but is covered in some detail elsewhere [5.8].

Summary

1 When studying a habitat and obtaining samples of the populations present, it is important to observe certain elementary principles of procedure.
2 A variety of nets are now available for all kinds of collecting. For the best results and the survival of the equipment, it is important to select the appropriate net for the purpose and to treat it kindly.
3 Animals inhabiting bushes and small trees can be beaten into a beating tray. An old umbrella is often an effective substitute.
4 Animal traps are of many kinds and include pitfall traps for soil arthropods, mammal traps, water traps for aquatic insects, and light traps for moths at night. All have their advantages and limitations.
5 Sampling soil populations presents particular difficulties. Chemical methods are effective for earthworms, while flotation and Tullgren and Baermann funnels are useful for arthropods and other small organisms.
6 Quadrats are the standard method for sampling plant populations. Before using them, consideration should be given to the elementary mathematical principles involved, particularly the problem of error.
7 Percentage cover is a useful parameter for measuring the success of different species of plants colonising the same habitat or of one species colonising different habitats. The point frame enables percentage cover of relatively small areas to be estimated with precision.
8 Belt and line transects can be used to study plant distribution over relatively large areas. The accumulation of data by these methods tends to be time-consuming.
9 Estimating the density of animal populations presents peculiar difficulties on account of their mobility. A variety of methods is available depending on the nature of the habitat and the size of the organisms. Procedures involving mark, release, and recapture are widely used for the study of populations of large animals, such as snails and flying insects, which are easy to handle and mark without damage and which randomise readily.

References

5.1 Leadley Brown, A. (1978) *Ecology of Soil Organisms*, Heinemann

5.2 Phillipson, J. (Ed) (1971) 'Methods of study in quantitative soil ecology: population, production and energy flow', *International Biological Programme, Handbook No. 18*, Blackwell

5.3 Heath, O. V. S. (1970) *Investigation by Experiment*, Studies in Biology No. 23, Arnold

5.4 Fisher, R. A. and Yates, F. (1967) *Statistical Tables for Biological, Agricultural and Medical Research*, Oliver and Boyd

5.5 Lindley, D. V. and Miller, J. C. P. (1953) *Cambridge Elementary Statistical Tables*, Cambridge

5.6 Ashby, M. (2nd edn. 1969) *An Introduction to Plant Ecology*, Macmillan

5.7 Greig-Smith, P. (1957) *Quantitative Plant Ecology*, Butterworth

5.8 Southwood, T. R. E. (2nd edn. 1978) *Ecological Methods*, Methuen

5.9 Yapp, W. B. (1957) 'The theory of line transects', *British Birds*, **3**, 94–104

6 Methods of study

Measuring the environment

Advances in technology have now made it possible to measure and record most aspects of the physical environment, but at the outset of any ecological investigation it is sometimes difficult to decide just what measurements are likely to be needed. For this reason there is a strong temptation, to which many students succumb, to record everything possible in the hope that some of the information will turn out to be useful. Where time is no object and there is an abundance of computer assistance, such a policy may be justified, although all too often it is merely the result of a lack of planning. In class studies and individual project work, an excessive emphasis on environmental measurements can lead to an outcome which is nearer climatology than ecology. When beginning an ecological study we therefore need to consider carefully *in advance* the kind of environmental factors that may play a predominant part in influencing the population or community concerned. Some measurements that are likely to be relevant in the study of different habitats are summarised in Table 6.1.

Table 6.1 Environmental measurements likely to be relevant in the study of different habitats

Habitat	Measurements likely to be relevant
Fresh water	Temperature, light, oxygen, pH, hardness
Brackish water; sea	Temperature, light, oxygen, pH, salinity
Open field; wood	Temperature, light, humidity
Soil	Temperature, pH, specific ions, humidity, organic content

Soil

Soil composition

The ecological factors in soil (**edaphic factors**) play a major part in determining the size and distribution of the plant and animal populations associated with it. Prominent among these factors are the nature of the rock particles in the soil and its content of such components as air, water, organic matter, and minerals. The soil particles are largely derived from rocks by weathering, the processes of erosion breaking down the rock surface into particles of varying size. Volcanic rocks such as granite are hard, tend to erode slowly, and form relatively large pieces. Softer sedimentary rocks such as sandstone wear more readily and produce smaller particles. Particle size plays an important part in determining the ecological characteristics of soil. In general, the larger the particles, the more easily drainage will occur, with the consequent loss of important mineral nutrients and bases. Many porous soils tend to become acid, as in heathland. Conversely, small particles restrict drainage. This is the case for clays, which, although tending to remain wet, conserve their mineral component and are therefore more fertile.

Obtaining a soil sample

In order to obtain a sample of soil from a particular depth, it is frequently inappropriate to dig a pit and a **soil auger** must be used instead. There are two forms:

(i) a helical type consisting of a metal rod about 650 mm long with a screw head of diameter about 40 mm and length 200 mm (Fig. 6.1(a)). At the other end of the rod is an eye for the insertion of a tommy bar. To obtain samples, the

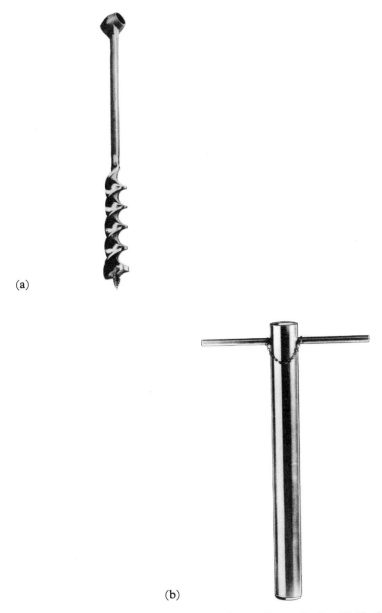

(a)

(b)

Fig. 6.1 Soil augers (a) helical type, (b) sampling cylinder (Philip Harris Ltd)

surface vegetation is removed from a small area, and the auger driven in to a distance of around 150 mm before being removed. The soil is laid out on a tray and a second sample then obtained from a similar depth below the first, and so on. The various samples can then be compared and a profile constructed. Deeper samples may become contaminated as they are extracted through the original bore hole, and in order to avoid this, an auger with a larger diameter should be used to begin with. The procedure is useful in establishing the nature of soil zonation but it does not provide an accurate picture of the extent of each zone.

(ii) for quantitative studies of soil populations it is usually necessary to remove a core which preserves the various soil components in their original relationship with one another. This requires a cylindrical type of auger in which the screw end described above is replaced by a steel sampling cylinder about 220 mm long and 40 mm in diameter (Fig. 6.1(b)). Alternatively, a satisfactory home-made version can be constructed from any suitable piece of metal tubing which has been sharpened at the leading edge (Fig. 6.2).

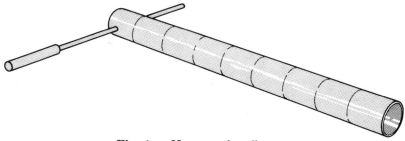

Fig. 6.2   Home made soil auger

When obtaining soil samples by either method, it should be rememberea tnat soil zonation can vary appreciably within a metre or so. It is therefore always wise to obtain several sets of samples from the same site for comparison. To ensure comparability and separation of samples, lines 150 mm apart should be drawn on the inspection tray and care taken to ensure that successive samples are laid out in order of depth.

Soil classification

Soil particles are defined arbitrarily by the size categories adopted by the International Society of Soil Science in 1927. These are summarised in Table 6.2.

The artificial separation of these various components is known as **mechanical analysis**. Gravel and sand are well suited to such treatment and can conveniently

Table 6.2   Categories of soil particles

Fraction	Diameter of particles (*mm*)
Gravel (grit)	>2.0
Coarse sand	2.0–0.2
Fine sand	0.2–0.02
Silt	0.02–0.002
Clay	<0.002

Fig. 6.3 Nesting aluminium sieves for the mechanical analysis of soil (Philip Harris Ltd)

be quantified by passing samples of dried soils through sieves of standard mesh (Fig. 6.3). The different fractions can be weighed and expressed as a percentage of the whole. Sets of analytical sieves are expensive, however, and if these are not available it is possible to use nests of yogurt pots in which the bottoms have been replaced by a sheet of tin foil. Fisher [6.1] has described the method in detail together with its practical applications. The biggest problem is piercing the tinfoil with holes of appropriate sizes. For gravel, metal knitting or sewing needles are suitable and their diameter can be measured with a micrometer. Finer sizes can be made with hard wire of the following guages: for coarse sand 25 swg, for fine sand 34 swg, and for silt 40 swg. The fineness of silt and clay particles cause difficulty when using this simple procedure and results tend to be inaccurate unless special precautions are taken [6.1].

Most elementary ecological studies demand no more than an approximate idea of the soil constituents present and this can easily be obtained by shaking a quantity of fresh soil with water in a tall vessel such as a measuring cylinder and allowing it to settle (Fig. 6.4). The various layers will be arranged according to

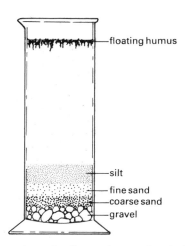

Fig. 6.4 Separation of soil constituents by shaking with water

particle size, the coarsest constituent (gravel) being at the bottom. Clay tends to form a murky suspension in water that takes some time to settle, but it can be made to flocculate quite quickly by the addition of an electrolyte such as the calcium in lime water. This, incidentally, is one of the reasons for adding lime to clayey soils, since the coagulation of clay particles makes the ground more porous and therefore improves drainage. Humus (semi-decomposed plant matter) is often less dense than water so much of it floats near the surface. However, this is not an accurate enough method of estimating the proportion present.

A typical analysis of the five commonest types of soil is shown in Table 6.3.

Table 6.3 Approximate analysis (%) of five common soils (after Hall)

	Sandy loam	Clay loam	Loam	Chalky	Peaty
Water	2.4	4.4	2.4	2.4	8.4
Humus	4.0	6.4	14.3	6.9	32.8
Coarse sand	10.3	3.9	0.1	20.2	26.6
Fine sand	67.0	30.1	53.3	21.9	14.2
Clay and silt	16.3	55.2	29.7	9.6	18.0

Water content

The wetness of soil and its capacity for retaining moisture are two of the most important physical factors determining the spatial distribution of plant species. The total water content of soil is the sum of the following constituents:

(i) water which drains from the surface downwards carrying valuable bases and minerals with it (**gravitational water**). This builds up after rain but declines once drainage has taken place.

(ii) water held within the pore spaces and passing between them by capillary action (**capillary water**). Both this and gravitational water are available to soil organisms. They can also be removed, at least in part, by heating.

(iii) imbibed water within fine clay particles which is retained by strong osmotic forces (**osmotic water**).

(iv) the thin film of water surrounding soil particles which is strongly adsorbed onto their surfaces (**hygroscopic water**).

Both (iii) and (iv) are much less available to plants than (i) and (ii).

The simplest method of estimating water content is to heat to constant weight in an oven at about 105°C a sample of, say, 10 g of fresh soil. The loss in weight expressed as a percentage will account for all the hygroscopic water and some of the capillary water. In order to avoid the temporary effects of gravitational water, it is advisable to allow a period of at least 24 hours to elapse after heavy rain. In most ecological studies, it is seldom necessary to obtain very accurate measurements of soil water content and the heating method described above will generally suffice for all comparative purposes.

Humus content

Soils contain varying amounts of semi-decomposed organic matter (**humus**) usually ranging between 5 and 15 per cent (wet weight). However, in peaty soils (**mor**) such as characterise heathland (p. 237), it may exceed 30 per cent (Table

6.3). The organic component of fine earth will be largely in the form of colloidal particles which can be regarded as humus for all practical purposes.

Humus is important for two main reasons:

(i) it retains moisture while aiding drainage. This is in contrast to clay which, although it also retains moisture, tends to inhibit drainage on account of the compacting of the fine particles and their readiness to swell with imbibed water.
(ii) it is gradually broken down in the soil by organisms of the carbon and nitrogen cycles, releasing minerals that are essential for plant growth.

As with soil water, precise estimates of humus content are difficult, but loss in weight on ignition gives a reasonable estimate of organic content which is adequate for comparative purposes. About 5 g of soil dried at 105°C and cooled in a desiccator are weighed in a crucible and heated to red heat for about half an hour. Cooling, weighing, and reheating are continued until no further loss occurs. The weight loss is then expressed as a percentage, which represents the amount of oxidisable organic matter present. One of the chief sources of error in alkaline soils (e.g. chalk and limestone) will be the decomposition of considerable amounts of inorganic minerals, particularly carbonates, with consequent loss of CO_2. To make a correction for this, add a little ammonium carbonate solution to the cooled solid and then heat in an oven at 105°C to drive off the excess liquid. Cool and reweigh. The gain in weight will be the amount of CO_2 initially lost through heating.

One of the main reasons why such huge populations of animals live in the soil is that its composition provides an insurance against the danger of desiccation. Thus, the water and humus content together are a reliable index for predicting the density of animal colonisation.

Clay content

As indicated earlier, the estimation of the clay content of soil presents certain difficulties. Freeland [6.2] has described a useful method of partly overcoming these problems by making measurements of turbidity using a colorimeter. Clay suspensions of known concentration are prepared and their absorbances plotted. Unknown samples of soil are then treated similarly. After the heavier particles of gravel and sand have settled, the absorbance of the remaining turbidity is estimated and compared with the clay standards. The procedure is relatively simple and works well. However, its precision rests on the assumption that the turbidity produced by a soil sample when shaken with water is entirely due to clay. In soils with a high organic content, such as garden loam, this is probably not so, and a proportion of the turbidity is likely to be due to fine particles of organic matter which remain in suspension. However, the procedure certainly has its merits and can be useful for making approximate comparisons.

Temperature

Variations in temperature not only govern the rate at which living organisms grow but also play an important part in influencing animal behaviour. As we saw in Chapter 2, the effects of temperature change are seldom exerted alone but usually act in combination with one or more other physical factors such as light. In ecological studies we need not only to measure temperature at a particular place and time, but also to note variations which may take the form of short-term gradients (**thermoclines**) as in ponds and the sea in summer, or more long-term fluctuations such as seasonal changes. For such purposes the ordinary mercury or alcohol thermometer has certain limitations in that it is easily broken and can only record the temperature at a particular moment (it cannot measure a thermo-cline). The maximum and minimum thermometer is little better and has the added disadvantage of being bulky and thus likely to destroy the gradient it is intended to measure.

Electronic thermometers

A variety of thermometers are now available which are relatively inexpensive and largely overcome the above difficulties. The sensors used are of three kinds:

(a) Thermistors
These are solid-state semiconductors which exhibit changes of resistance with temperature. Since the changes are non-linear, the scales used by manufacturers tend to be arbitrary. These are good enough for comparative purposes, but if actual readings of temperature are required, calibration must be carried out by the user. As Crellin [6.3] has pointed out, a complication can arise due to the fact that the response of a thermistor thermometer may vary with the thermal conductivity of the medium in which the measurements are made. A separate calibration may therefore be needed for each medium. However, elementary ecological studies are usually based on relatively few different habitats so this problem should not be insuperable.

Thermistors are made in all sizes and shapes and can easily be mounted on probes for submersion in water, burying in the soil or use in a variety of situations on land. Moreover, thermometers incorporating them are easily home made using a simple Wheatstone Bridge circuit as illustrated in Fig. 6.5. R_1 and R_2 are fixed resistances, while R_3 is variable to counteract changes in the thermistor, and can be calibrated with a temperature scale (or set of numbers for conversion on a graph). G is a sensitive galvanometer which indicates when the two sides of the bridge are balanced. Experience has shown that this type of thermometer is hardwearing and reliable. Moreover, the electrical characteristics of thermistors do not change appreciably with time.

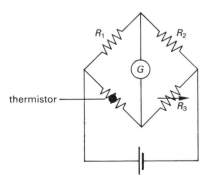

Fig. 6.5 Principle of the resistance thermometer using a thermistor in a Wheatstone Bridge circuit

(b) Semiconductor diodes

These are based on the principle that if a continuous current is passed through a silicon diode, the potential difference across it is almost directly proportional to the temperature of the diode. This type of sensor is not subject to the kinds of problems occurring with thermistors and its sensitivity is sufficient for most ecological purposes. It provides the basis for most commercial meters.

(c) Thermocouples

These are devices which make use of the principle that a potential difference is directly proportional to the difference between two heat sources. The simplest way of obtaining an absolute reading is therefore to maintain the cold source at a constant temperature. Several types of instrument based on this principle are available commercially and they are particularly useful for measuring temperature changes. They are, however, less sensitive for temperature measurement and therefore not so suitable for elementary ecology.

Several commercial electronic thermometers are available, such as that shown in Fig. 6.6 which is based on semiconductor diodes and is calibrated to give direct readings in °C. Many models are supplied with a variety of probes suitable for different media and giving a wide range of temperatures from which to choose. For ecological work it is unlikely that a range of say $-5°C$ to $+40°C$ will ever be exceeded, a point to be borne in mind when purchasing a thermometer. For many investigations, all that is required is a quantitative comparison of one situation with another, so arbitrary numerical values will suffice. For this purpose the environmental comparator using thermistors (Fig. 6.7) has proved useful, but the scale is empirical and calibration is needed if absolute readings are required.

Fig. 6.6 Battery operated resistance thermometer (Philip Harris Ltd)

Fig. 6.7 Environmental comparator (Griffin and George Ltd)

Light

Of all the physical factors with which we are concerned in the study of ecology, light holds pride of place because of the number, diversity, and importance of its effects on both plants and animals. However, there are considerable problems in measuring it, not only because of the complexity of light itself, but because the level of solar illumination fluctuates so greatly from one moment to the next. Also, we must remember that, like most physical factors, light seldom exerts its effects alone but acts in conjunction with other factors, particularly temperature.

One of the most powerful influences on plant distribution in relatively calm waters such as ponds, lakes, and the shallow sea, is turbidity due to floating matter. This reduces the penetration of light and hence the level of photosynthesis of the plants below. For comparative purposes, estimates of visibility can be useful and these are easily obtained by means of a **Secchi disc** (Fig. 6.8). This consists of a

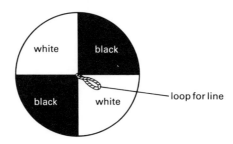

Fig. 6.8   Secchi disc for comparative studies of visibility in water

circular plate of wood, metal or plastic, about 200 mm in diameter, painted with opposing black and white quarters, and weighted centrally to ensure that it floats horizontally when submerged. It is attached to an appropriately calibrated line, either by a loop at the centre or, for greater stability, at three points round the edge. To determine 'Secchi disc visibility' the disc is lowered *slowly* into the water until it just disappears from view, when the depth is noted. It is then lowered by about another 0.5 m and slowly raised until it reappears, this depth also being recorded. The average of the two readings is the Secchi disc visibility. It is important to carry out the procedure slowly, partly to ensure the most accurate measurements of depths of disappearance and reappearance, and partly to avoid, as far as possible, upsetting the environmental conditions that we are attempting to study.

Electronic light meters

The measurement of light in ecological situations is a good deal more complex than appears at first sight. There are two particular reasons for this:

(i) the spectral response of the instrument must be comparable with that of the organisms studied. These are frequently plants, which absorb mainly blue, red, and infra-red radiation.

(ii) photosensors of the kinds usually used in light meters do not show a linear response to changing illumination.

Sensors are of two kinds:

(a) Photoconductors

These are of the cadmium sulphide type which are widely used in photographic light meters and whose resistance varies with the amount of light falling on them. They are particularly sensitive to red and infra-red light. The relationship between incident light and resistance is non-linear but with suitable circuitry it can be made logarithmic as in photographic light meters.

(b) Photovoltaic cells

These are of the selenium type and generate a potential difference when light falls on them, the spectral response being somewhat similar to the human eye. The output of the cell is nearly linear, which explains why it is preferred in calibrated light meters. One problem with selenium cells is that their output decreases with age and they therefore need to be recalibrated from time to time.

Calibration of light meters

The amount of *visible* radiation falling on an area (**illumination**) is measured in lux (lm m^{-2} or lx). The *total* radiation falling on an area (**irradiance**) is measured in watts per square metre (Wm^{-2}). Since plants utilise portions of the spectrum to which the human eye is least sensitive (blue, far red and infra-red), ecologists tend to use Wm^{-2} (or similar units) even though these refer to the total radiation rather than just to particular wavelengths.

For the reasons outlined above, many light meters designed for elementary ecology are uncalibrated and use only arbitrary scales. These are quite adequate for comparing illumination (or irradiance) which is frequently all that is needed. They are often provided with probes of different lengths so that they can be used on land and under water. Crellin [6.3] has suggested a useful method for calibrating a light meter with sufficient accuracy for most ecological needs using a 100 watt electric light bulb, and this is reproduced in Table 6.4.

Table 6.4 Data for calibrating light meters. Light values in lux, can be converted approximately to Wm^{-2} by dividing by 50

Distance from 100 watt, clear, 240 VAC single-coil bulb (cm)	Illumination (lx)
10	7500
15	3200
20	2000
30	900
40	550
50	350
60	250
80	150
100	100
150	45
200	25

The use of light meters

A further note of caution is needed regarding the use of light meters and the interpretation of information obtained from them. Even on an apparently cloudless summer day, the level of illumination at a particular site can vary considerably in a matter of seconds. So if comparative measurements are to mean anything, they should be taken simultaneously. This means that two (or more) meters will be required. To avoid excessive cost, these can easily be constructed in any reasonably equipped laboratory [6.4]. Even if we manage to achieve momentary comparability, there are still limitations to the information obtained if we are aiming to interpret problems such as patterns of plant distribution. These will depend not on temporary fluctuations in illumination, but on the total amount of assimilable light energy falling on the plants over the period of time that their green parts are active. To measure this parameter requires a cumulative type of light meter or **integrating photometer**. A number of designs exist, but most include a voltameter linked to a photosensor such as a selenium cell, which, when it emits a current, records total irradiation in terms of the deposition of metal on an electrode. A particularly convenient form makes use of silver electrodes and an electrolyte consisting of silver nitrate solution and acetic acid contained in plastic Weedol tubes. The design of the apparatus has been described by Westlake and Dawson [6.5].

An alternative approach to the same problem has been described by Bishop [6.6] who has designed an integrating photometer based on relatively cheap integrated circuits. It consists essentially of four parts:

  (i) a generator which produces pulses at a rate proportional to the incident light
  (ii) a 'latch' controlling the pulses passing from the generator to the counter
 (iii) a clock governing the latch and the frequency of the passage of pulses
 (iv) a counter registering the number of pulses it receives.

Among the advantages claimed for the equipment, which could be constructed in any suitably equipped workshop, are low cost, portability, robustness, high sensitivity, and the facility for instantaneous readings. In many respects, it is therefore preferable to the photovoltameter.

Humidity

There are various measurements that we can make of the amount of water in the atmosphere, some being of more use in the study of ecology than others. We can determine its **relative humidity**, which is defined as the ratio of the mass of water vapour per unit volume of air to that of a similar volume of saturated air at the same temperature, expressed as a percentage. Again, we may be concerned with its **evaporating power** which is a function of wind speed, temperature, and relative humidity. Perhaps a more useful measure is the **saturation deficit** of the air – the amount by which the partial pressure of water vapour falls short of the partial pressure at saturation point irrespective of temperature.

A variety of instruments exists for the measurement of atmospheric humidity. These instruments are usually known as **hygrometers**. The wet and dry bulb type occurs in both a hanging and a hand-held whirling form, and there are several designs of hair hygrometers. Paper hygrometers are sometimes used but their rate of response is slow and their accuracy low. Again, potometers and atmometers measure evaporation either from the surface of plants or from the moist surface of synthetic media such as porous pots. The main disadvantage of all these methods is that they only measure the **macroclimate** which, as we now know, usually differs substantially from the conditions in which plants and animals actually live (**microclimate**). Thus the ecological study of butterfly larvae inhabiting grassland has shown that the climate one metre above the ground is no guide to the environment within a grass tuft immediately below [6.7].

One of the problems of measuring the humidity of microclimates is to ensure that the procedure used does not itself alter the environmental conditions it is intended to measure. For instance, in determining the humidity of air in grass tufts a standard pattern of wet and dry bulb thermometer would obviously be inappropriate. Cryptozoic animals such as woodlice, centipedes, and millipedes are particularly sensitive to changes in moisture (also to the influence of temperature and light), so in studying them we may need to make fairly accurate measurements of their physical environment. The best that can be done at present is to use the cobalt chloride method [6.8]. Cobalt chloride paper is prepared by dipping a filter paper in a 25 per cent aqueous solution of cobalt chloride ($CoCl_2.6H_2O$) and allowing it to dry. It is then cut up into strips which will fit conveniently between two microscope slides. There is a close relationship between colour and atmospheric humidity, the paper being blue at low and pale pink at high humidities, with a series of lilac colours in between. To measure the moisture of a microhabitat, one of more papers are inserted, each between two microscope slides to protect them from dirt and water. The length of exposure needed varies somewhat, depending on temperature and the relative humidity of the atmosphere, but two hours is normally sufficient.

Standards for comparison can be prepared by exposing pieces of cobalt chloride paper to different humidities measured by a hair hygrometer in a tank such as an aquarium. The coloured papers are then sealed in liquid paraffin between two microscope slides with wax or Canada balsam. Relative humidities ranging from 40–70 per cent can be measured by this means to an accuracy of 2 per cent. Beyond this range the error increases to around 5 per cent. For higher humidities cobalt thiocyanate is preferable to cobalt chloride: it has the added advantage that equilibrium is usually reached in about half an hour. A humidity testing kit is produced commercially using cobalt thiocyanate papers and a set of coloured glass standards for comparison [6.9]. However, the fact remains that there is no satisfactory method of measuring humidity in the field. A portable electronic meter and humidity probe is badly needed.

Current

Water currents in habitats such as streams and rivers play an important part in influencing plant and animal distribution. They are not easy to measure, partly because of the difficulty of designing a good flow meter cheaply and also because natural waters seldom flow smoothly, but are liable to varying amounts of turbulence resulting in considerable fluctuation in the readings obtained. For elementary comparative studies, three methods of measuring current are available:

(i) observation and timing of a light floating object over a given distance – a table tennis ball is ideal for the purpose. This method has the limitation of

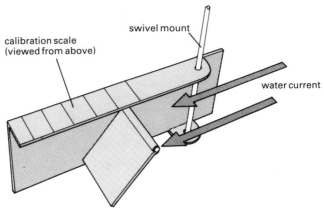

Fig. 6.9 Deflection current meter. Its sensitivity depends upon the length and breadth of the hanging vane (after Bennett and Humphries)

measuring only the rate of flow at the surface; it gives no indication of water speed deeper down, which is where many plants and animals live.

(ii) deflection of a hanging weighted vane from the vertical (Fig. 6.9), the sensitivity of the instrument depending on the length and breadth of the vane. Calibration is carried out at the surface by method (i) above. It is important that the markings should be easily visible when submerged. It is a crude device but can be useful if quick approximations are required for comparative purposes.

A much more precise instrument using the same principle can be devised as follows. If a single vane is attached to a light metal rod which is held vertically (rather like the rudder of a boat) and is allowed to swing freely in a current, it will position itself pointing downstream. If a second vane is attached above the first and at an angle to it (see Fig. 6.10) the pressure of the water on the outer surface will be greater than on the inner and a deflection will occur. The magnitude of the deflection will be related to the rate of flow. An appropriate indicator and scale can be attached to the top of the moving rod and calibrated in m s^{-1}. A commercial version known as a Flowvane is available [6.10]. It is 93 cm long, weighs 0.8 kg and can be submerged to a depth of 60 cm. The velocity range covered is stated to be 0.15–2.5 m s^{-1}. The apparatus has the obvious advantages of simplicity, portability, and the absence of parts needing replacement such as batteries. A satisfactory version could be constructed cheaply in any workshop, the chief factors influencing its sensitivity being the lightness of the moving parts and their lack of friction against supporting surfaces.

Fig. 6.10 The 'Flowvane' stream flow meter

(iii) the reversed L-tube type of meter (Pitot tube) was designed by the author many years ago and in spite of its imperfections, has stood the test of time. It depends upon the principle that if two L-shaped tubes are mounted back to back and arranged so that one points upstream and the other down, a positive pressure will be built up in the former and a negative one in the latter. The difference between the two can be displayed and measured if each is connected to opposite sides of a manometer (Fig. 6.11). To avoid breakage, the lower parts of the tubes should preferably be made of metal, for example copper, and should be as wide as is reasonably possible. The manometer contains water coloured with a dye such as borax carmine or eosine. The current speed is determined by the *difference in height* of the two columns, which can be read off on a piece of graph paper mounted behind them. Calibration is carried out by the floating object method already described (see (i) above) and a graph prepared converting deflection in millimetres to metres per second. This can be mounted on the back of the board and protected with a coating of polyurethane varnish. In practice, it is advisable to fit clips at the two ends of the manometer to avoid spillage of liquid when not in use. The apparatus is inevitably rather cumbersome to carry about, for if parallax errors in reading the manometer are to be avoided, the board must be about one metre long. This difficulty can be partially overcome by hinging the apparatus in the middle.

Summarizing the disadvantages and advantages of this simple piece of equipment:

Disadvantages
(i) rather cumbersome to transport
(ii) liquid in the manometer tube tends to oscillate due to water turbulence, so an average reading is needed

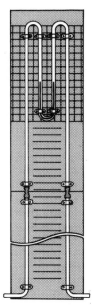

Fig. 6.11 Current meter based on the reversed L-tube principle

(iii) a tendency for air-locks to develop in the lower part of the L-tubes. These can readily be detected by the abnormal manometer readings they produce and can usually be removed by tapping the apparatus at the bottom

Advantages
 (i) more accurate than the other methods described
 (ii) easily made in any laboratory
(iii) demonstrates beautifully the decreasing speed of the current with increasing depth of water. On a stream bed there is virtually no current and this explains why weakly swimming organisms such as the freshwater shrimp, *Gammarus pulex*, and numerous small mayfly nymphs are able to survive there without being washed away.

Hydrogen ion concentration (pH)

Mention was made earlier of **indicator species**, plants and animals which, by their presence in a habitat, provide a clue to the nature of the environment. Among plants we have **calcicole** species which can only flourish on chalk and limestone soils where the pH is on the alkaline side – usually around 7.5–8. By contrast **calcifuges** cannot exist on calcareous soils, typical examples being the inhabitants of heathland, where the pH of the soil water is frequently about 4.5. Hydrogen ion concentration is thus a valuable indicator of the species of plants and animals likely to be found in a habitat.

The quickest way of making an approximate estimate of pH is by means of a multiple indicator such as the BDH soil indicator, which covers a pH range of 4–8, the colour changes corresponding to steps of 0.5 pH. Various kits, which include colour charts for matching, are now supplied by British Drug Houses Ltd, the most accurate of them providing readings to the nearest 0.2 pH. In studies of soil, one of the problems of using indicators is that suspended clay particles tend to obscure the colour of the indicator. This difficulty can be overcome by adding a little barium sulphate, which causes the clay to flocculate without affecting the pH.

Electronic determination of pH

For a more accurate and quicker estimation of pH under field conditions, various small, portable, battery-operated meters are now available commercially. A typical example is shown in Fig. 6.12. These are supplied with a variety of electrodes and can be calibrated using standard buffer solution to give a pH range of 0–14.

There are various precautions that need to be taken when using these instruments in the field. The electrodes are fragile and must be treated with care; they should be kept immersed in distilled water when not in use. For studies of soil, special spear electrodes have been developed, which will withstand pushing from above and friction from the sides. Again, electrodes need standardising with a buffer solution of known pH, as near as possible to the time when they will be used. A slight contamination can alter the pH reading and this underlines the importance of careful cleaning with distilled water before a measuremenet is taken.

Fig. 6.12 Battery operated pH meter for use in the field (Griffin and George Ltd)

Sampling

When considering the adaptations displayed by living organisms in relation to their habitats, we tend to think of them as adjustments to a diverse physical environment – and so they are. But in the process of adjustment, plants and animals sometimes bring about radical changes in the environment itself, a principle which is well illustrated by the most important of all the dissolved gases, oxygen.

Before we can make an estimate of the substances dissolved in water, it is necessary to obtain a representative sample from the locality we wish to test, which

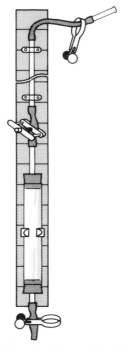

Fig. 6.13 Sampler for use in shallow water

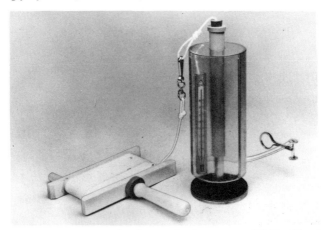

Fig. 6.14 Water sampler for use in deep water (Philip Harris Ltd)

may be some way below the surface. Sampling in shallow water presents little difficulty and can be done by hand. All that is needed is a bottle with a tightly fitting stopper which can be removed under water, the bottle being allowed to fill on its own. The stopper is then replaced, care being taken to ensure that no air bubbles are trapped inside.

For depths up to about two metres an apparatus such as that shown in Fig. 6.13 can be constructed. This consists essentially of a calibrated rod to which a glass tube container fitted with corks is joined at one end, plus tubes which can be clipped. A flexible tube leads upwards to a mouthpiece which is also provided with a spring clip. When constructing the apparatus, it is important to ensure that the ends of the two glass tubes entering the container are flush with the corks, otherwise air bubbles will be trapped. To obtain a sample, lower the apparatus to the required depth with the clips of the water container removed. Suck until the container is full and water has passed some way up the tube towards the mouthpiece. Clip the mouthpiece and remove the apparatus from the water. Apply the clips on either side of the water container, then remove the container.

For sampling at greater depths the above apparatus is too cumbersome and a different design is required. Several types are produced commercially, a good example being that shown in Fig. 6.14. To take a sample, the apparatus is lowered to the required depth as indicated by the markers on the line. The floor of the container is then closed by a sharp tug on the line, a good seal being ensured by two washers, one rubber and one foam. As the sampler is retrieved, the water pressure closes the top valve, trapping the sample. When required, the water sample can be drained by standing the sampler on a level surface and opening the tap of the drain tube at the base of the cylinder. A wooden reel is provided, with 10 m of line marked with tags at 0.5 m intervals.

Estimation of oxygen

There are several reasons why students of ecology may wish to gain some idea of the level of oxygen dissolved in water. Reference has already been made to the influence of plants on their environment. On a bright summer day the amount of oxygen at the surface of a well-illuminated pond will frequently exceed saturation (i.e. it will contain actual bubbles of oxygen). This will also be true of the shallow waters of streams, but here aeration is largely due to turbulence. The availability of oxygen relates closely to the development of gills and other respiratory organs in the young and adult stages of many small aquatic arthropods. By contrast, some burrowers in mud, such as the larvae of some Diptera (for example the midge *Chironomus*) make use of respiratory pigments like haemoglobin which aid oxygen uptake, indicating a degree of adaptation to anaerobic conditions.

The amount of oxygen (or any other gas) present in water can be expressed either as parts per million, as the volume at STP ($cm^3 dm^{-3}$) or as a percentage of air saturation. The last represents the oxygen present as a fraction of the amount dissolved in water at equilibrium with air at the same temperature. As indicated earlier, in shallow water with much vegetation this figure often exceeds 100 per cent during the summer.

(a) Chemical estimation

The commercial development of battery-operated oxygen meters has greatly assisted studies of water oxygenation. However, there are still occasions when it may be convenient to collect a number of samples for comparison, and estimate their oxygen concentration chemically. Two different procedures are available. In the first the water sample is made alkaline and titrated with iron (II) sulphate solution. Any oxygen present converts the iron (II) to iron (III), the end point being determined by a redox indicator. In practice, the procedure has proved rather erratic and the degree of repeatability is low. Moreover, the need to carry out titration at once makes it rather unsuitable for use in the field.

The standard Winkler method is well tried and involves the addition to the water sample of a manganese (II) salt. The solution is made alkaline. The manganese (II) is oxidised to manganese (III) and in this state the oxygen in the water is 'fixed'. Final estimation can be carried out later. Upon acidification in the presence of potassium iodide, iodine is released in an amount equivalent to the dissolved oxygen present. This can be determined by titration with sodium thiosulphate (VI) using starch solution as an indicator. Gill [6.11] has devised an excellent version of the Winkler procedure using plastic syringes, which experience has shown to be well suited for work either in the field or in the laboratory. His technique, together with the necessary apparatus and reagents, is summarised as Appendix A at the end of this chapter.

(b) Electronic estimation of oxygen

The commercial development of small portable dissolved-oxygen meters has enabled measurements to be made quickly in the field and with sufficient accuracy for elementary ecological purposes. However, the preparation and use of these instruments is by no means as straightforward as a pH meter, and various precautions have to be taken if reliable results are to be obtained. Richardson [6.12] has provided a valuable review of oxygen electrodes and their uses together with suggestions for the construction of home-made circuitry to go with them. For a detailed discussion of the problems relating to their use reference should be made to this account.

Sensors for detecting dissolved oxygen in a sample of water are of two kinds. Galvanic probes consist of a pair of dissimilar metals mounted behind a membrane to form a cell. The current flowing between the electrodes is controlled by the rate at which oxygen diffuses inwards, destroying by oxidation the layer of hydrogen that builds up on the cathode. The most usual arrangement is a lead anode and a silver cathode surrounded by an alkaline medium, the amount of oxygen present being a function of the current produced. Polarographic probes are not cells but a polarizing voltage is applied between the two electrodes, which are usually both silver in a potassium chloride electrolyte. Since the porosity of the plastic film covering the electrodes varies with temperature, the sensitivity of the instrument can vary appreciably.

A typical commercial instrument is shown in Fig. 6.15(a). This consists of a

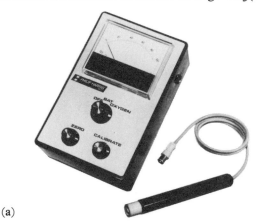

(a)

(b)

Fig. 6.15 (a) Dissolved oxygen meter for use in the field (Philip Harris Ltd), (b) meter in use

galvanic probe on an extended lead, in which a silver/lead electrode is covered with a plastic (PTFE) film which is permeable to oxygen and uses 1 M potassium hydroxide as electrolyte. Since the response of these instruments is subject to so many individual variables such as temperature and cleanliness of the electrode, they are not supplied ready-calibrated. For the estimations of oxygen in water which are usually required, the easiest method of calibration is to prepare some air-saturated water by shaking it in a stoppered flask for at least a minute. The instrument is then set either to 100 per cent saturation or to the known oxygen concentration of air-saturated water at that temperature in mg dm^{-3} (see Table 6.5 p. 126).

(c) Estimating oxygen in sand and mud
Investigation of the level of anaerobic conditions in sand and mud presents obvious problems but they can be overcome to some extent by the use of unglazed earthenware tubes. These are filled with stream water, corked, and sunk in the mud for a period of about a week, so that typical conditions are restored and equilibrium can be reached between dissolved oxygen outside and inside. The tubes are then removed and the concentration of dissolved oxygen estimated by one of the methods described above. This can be compared with that of the surrounding water. Mud dwelling species often exhibit a distinct vertical zonation, so when studying them it is worth sinking tubes to a series of different depths.

Multi-purpose electronic monitors

As we have seen in the preceding pages, a variety of electronic instruments are now available for monitoring different aspects of the environment. So far we have been concerned with the measurement of individual factors, but manufacturers have taken the opportunity of developing multi-purpose kits in which the functions are unified to some degree. These have been reviewed by Crellin and Tranter [6.13]. Mention has already been made (p. 105 Fig. 6.7) of the environmental comparator which combines the measurement of temperature and light. More elaborate kits are of two basic kinds:

(i) a container which merely houses a number of separate meters performing different functions
(ii) a kit with only one or two meter/power units to which a range of meter modules can be connected.

The first is the more expensive design but has the advantage that a number of separate measurements can be made simultaneously in different places. The second design is not so flexible but a reduction in the number of meter/power units means a saving in cost.

Multi-channel data recorders

It was mentioned earlier in connection with illumination (p. 108) that if we wish to measure the total light falling on an area of vegetation we must have recourse to a continuously recording instrument. The idea can be taken further in the form of a multi-channel data memory, which is now available commercially, to include not only light but temperature, oxygen, and pH. The availability of such recorders opens up many possibilities in the field of project work, such as the study of cyclical environmental changes occurring in different kinds of habitats and their influence on the plants and animals that colonise them.

Estimation of dissolved minerals: chlorides

The natural waters that occur both around our coasts and inland contain large amounts of dissolved minerals. Some of these occur naturally, others are the outcome of human activities. Many of the latter solutes are now regarded as pollutants.

A wide range of devices and methods now exists for measuring such factors as noise levels in air and chemical and bacteriological changes in soil, air, and water. Such aspects of the environment can form an important component of ecological projects. For further information on these aspects of environmental monitoring reference should be made to the review by Crellin and Tranter [6.13].

The occurrence of chlorides is, however, a special case, as their presence and fluctuations provide important ecological characteristics of sea shores and estuaries. Sea water varies little in composition and for all practical purposes can be regarded as containing 3.1 per cent sodium and magnesium chlorides. In estuaries, fluctuations in salinity may be considerable as the tide ebbs and flows. Such variations often exert profound effects on the animal and plant populations colonising the estuarine zone which may have to tolerate a chloride range from undiluted sea water to almost freshwater conditions. The influence of these changes on behaviour patterns and spatial distribution will be discussed further in Chapter 11.

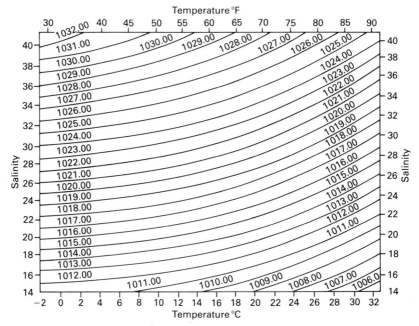

Fig. 6.16 Graphs showing the relationship between temperature, density, and salinity of sea water (based on former BA Chart No. C6104, with the sanction of the Controller, H.M. Stationery Office and of the Hydrographer of the Navy)

The density of sea water is a function of its temperature and salinity. An approximate estimate of the total chlorides present in a sample can therefore be made with a thermometer and sensitive hydrometer, using the conversion graph in Fig. 6.16. This simple method may well be precise enough for many kinds of study. For greater accuracy, dissolved chloride can be estimated chemically by titration with silver nitrate using potassium chromate as indicator. The procedure is as follows:

(i) Prepare an aqueous solution of silver nitrate containing $27.25 \, g \, dm^{-3}$.
(ii) Take a $10 \, cm^3$ sample of saline water and titrate with the silver nitrate solution using potassium chromate solution as indicator. The end point is reached when a permanent faint red colour of silver chromate appears.
(iii) The volume of silver nitrate required (cm^3) is approximately equal to the salinity in $g \, kg^{-1}$.

In theory, a small correction needs to be made (obtainable from tables) to allow for the slight variations in weight of a unit volume of sea water at different salinities. However, the overall limitations of salinity measurements discussed below hardly justify such refinement.

Electronic measurement of conductivity

One of the objections to using chloride estimation as a basis for judging salinity is that it assumes a constant relationship between chlorine ions and all the other ions present, no matter where the water originates. Such an assumption is obviously unjustified, so measurements of salinity have now been replaced by those of conductivity, which relates to the total ionic composition and not just that of chloride. Various commercial conductivity meters are now available, which are compact, easy to operate, and suitable for use in the field. Some have facilities for temperature compensation and all can be used in soil as well as in water. Their range covers fresh water with a conductivity of around 10^{-4} to $10^{-2} \, \Omega^{-1} \, cm^{-1}$ and sea water which reaches about $10^{-1} \, \Omega^{-1} \, cm^{-1}$.

Conductivity electrodes are quite easy to construct and Lapworth [6.14] has described a home-made version which could be easily and cheaply made in a science workshop. It should be emphasised, however, that while the title of his article suggests that his instrument measures salinity, in fact, like all conductivity meters, it is incapable of distinguishing between one type of ion and another.

Hardness of water

Another significant aspect of the ionic composition of water is its so-called 'hardness'. This can be of two kinds: **permanent**, caused by chlorides and sulphates of calcium and magnesium, and **temporary**, due to their respective bicarbonates. As we have seen (p. 16) the concentration of calcium (Ca^{2+}) and magnesium (Mg^{2+}) ions can play an important part in influencing spatial distribution, such as that of the water louse, *Asellus*, in the Lake District. The occurrence of calcium and magnesium ions is closely related to pH, for where they occur conditions will tend to be alkaline. Their estimation is unlikely to form part of most ecological studies, but if required, details can easily be obtained from the relevant literature [6.15].

Surveying methods

The study of most inland plant and animal communities usually involves sampling in a horizontal plane, so the use of procedures such as line and belt transects (p. 89) is appropriate. But on the sea shore where the major factor controlling the distribution of living organisms is the rise and fall of the tide, zonation is essentially vertical rather than horizontal. If we wish to assess and compare the situations at different levels on a shore, a means of measuring vertical distance is required. This is known as **surveying**.

The process of surveying takes place in steps and is illustrated in Fig. 6.17. The equipment required consists of a surveyor's level (Abney level) and two levelling staffs (or ranging rods) which are usually painted alternately black (or red) and white at about 0.5 m intervals. The level is essentially a device incorporating a sighting mechanism and a spirit level to ensure that it is horizontal when measurements are taken. The observer at A holds the level against a staff, preferably at one of the black and white junctions for ease of measurement. He then aligns his sight on the second staff held by his assistant, making sure that the level is horizontal. The assistant moves his finger up and down his staff until it is judged by the observer to be in line with the sights of the level. In Fig. 6.17, the difference between the readings on the two ranging rods is 0.7 m, indicating the height of point B above A. All surveying requires a point of reference; this is the line extending horizontally to the left from A and known as the **datum level**.

Abney levels tend to be expensive, and numerous home-made versions have been designed which serve the same purpose. A typical example is that devised by Nelson-Smith, which he has named the 'cross staff' (Fig. 6.18). The horizontal member is a wooden batten about 0.3 m long, into which a small spirit level is inserted. The horizontal member is secured between two vertical members at a right angle, forming a cross. A piece of mirror is fitted into slots on the inner surface of each vertical member so that it lies over the levelling bubble at 45°. A gap of about 20 mm between the lower surface of the mirror and the upper edge of the horizontal member enables the observer to sight the equipment and at the same time to see the levelling bubble reflected in the mirror. To increase accuracy, a fore-sight is attached to one end of the horizontal arm and a rear-sight to the

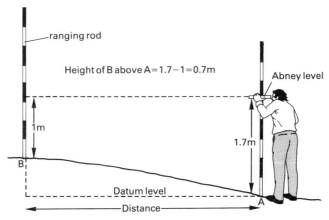

Fig. 6.17 Method of surveying sloping ground

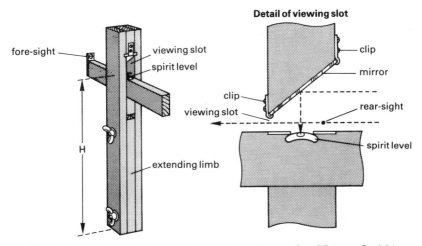

Fig. 6.18 Cross staff design of surveying level (after Nelson-Smith)

other (Fig. 6.18), which are just clear of the line of sight of the bubble. The length *H* from the rear sight to the base of the limb represents the vertical distance from one transect station to the next. Between the two vertical members is an extendable leg and when expanded to a maximum *H* is one metre.

To allow tidal levels to be included on the transect record and to permit comparisons between different shores, absolute levels are obtained by reference to predicted low water. Thus in Fig. 6.19, this is assumed to be approximately 0.8 m above Chart Datum. Predicted times and heights of low and high water for different parts of Britain can be obtained from Admiralty Tide Tables.

Having surveyed the shore to a point where a study of the plants and animals can begin, the next step is to set up a belt transect. This could be, say, four metres wide extending two metres either side of the line of the stations and vertically about 0.5 metres above and below the station marker (as shown in Fig. 6.19 at the three metre level).

A neat way of avoiding the tedium of surveying procedure and, incidentally, of achieving greater accuracy, has been described by Wood-Robinson [6.16] and is

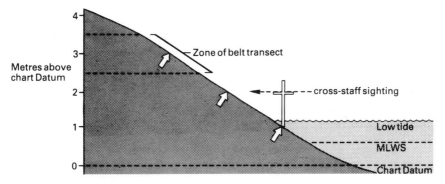

Fig. 6.19 Use of the cross staff for surveying a sloping shore. Chart Datum is the level used for Admiralty Surveys. MLWS is mean low-water springs (after Nelson-Smith)

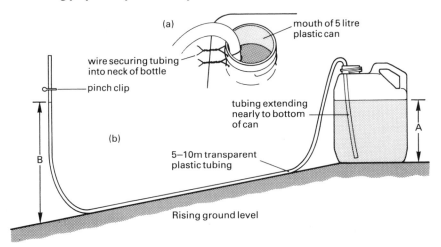

Fig. 6.20 Debenham Level method of surveying a rocky shore: (a) method of securing plastic tube in a water bottle, (b) using the level to survey a slope (after Wood-Robinson)

called the **Debenham Level**. The only equipment required is a translucent plastic can of about 5 dm³ capacity and approximately ten metres of clear plastic tubing. One end of the tubing is secured in the neck of the can as shown in Fig. 6.20(a), the other is closed by a clip. The can is filled with water, which is siphoned over into the tubing after removing the clip, and allowed to run to within about 50 cm of the end. The clip is then fastened again. The water level in the can is now adjusted to a mark on the outside, say at the 20 cm level (A in Fig. 6.20(b)). To measure the vertical distance between two sites, the can is placed at the upper site and the clipped end of the tube taken to the lower one. The clipped end is held vertically at a height slightly greater than that of the can and the clip is opened. The water in the tube will then settle at the same level as that in the can and its vertical height above ground can be measured (B in Fig. 6.20(b)). The vertical distance between the two sites will be B − A. This simple procedure is suitable for surveying many other kinds of ecological habitats, moreover, it has the advantage of cheapness and portability.

In conclusion, it is worth reiterating the point made earlier (p. 98) that determining the distribution and abundance of plant and animal species in densely populated habitats such as rocky shores can be a tedious and time-consuming undertaking. It can also lead all too easily to the accumulation of an incomprehensible mass of information. There is, therefore, much to be said for concentrating initially on a few 'indicator' species and studying them thoroughly before extending coverage later as circumstances and inclination dictate.

Summary

1   Although it is now possible to measure most components of the physical environment, it is important to decide which ones are relevant to a particular ecological enquiry.

2 When obtaining soil samples it is important to remember that zonation can vary appreciably within short distances.

3 The determination of the water and organic content of soils is less easy than it appears. Approximate values are useful and provide an idea for predicting the density of animal colonisation.

4 Electronic thermometers using thermistors or semiconductor diodes provide a sensitive means of measuring temperature and its variations.

5 Light meters based on photoconductors and photovoltaic cells have considerable limitations in the measurement of illumination. Many commercial versions require calibration if actual measurements are required. For long term studies, an integrating photometer is necessary.

6 Humidity measurements are difficult to make but important in revealing differences between microclimates and the surrounding macroclimate. Existing procedures are inaccurate and no satisfactory electronic instrument is available for use in the field.

7 Water currents can be measured in a variety of ways but the reverse L-tube method is most effective in demonstrating and assessing changes with depth. On the bed of a stream the water is almost still and this has important ecological implications.

8 The pH of soil and water can be significant for the survival of calcicole and calcifuge plants. It can easily be measured, either approximately with indicators or more precisely with an electronic meter.

9 Oxygen concentration in water plays an important part in animal colonisation. This can be measured chemically or electronically. Oxygen probes require careful treatment and calibration before use.

10 Multi-purpose electronic monitors are now available commercially enabling several different measurements to be made at once. Multi-channel data recorders can accumulate information on such variables as light, temperature, oxygen, and pH over a period of time.

11 The ions of solutes in water are not easy to estimate separately and for many purposes it is best to assess them collectively by measuring conductivity. Good electronic instruments are available for this purpose and can be home made.

12 The principles of surveying are well established and various home-made devices can be used, thereby reducing cost. For rocky shores, a siphon technique has much to recommend it.

References

6.1 Fisher, J. A. (1982) 'An elementary quantitative examination of sedimentary deposits', *School Science Review*, **64**, *No. 227*, 288–93

6.2 Freeland, P. W. (1975) 'Some applications of colorimeters and light meters in biology teaching', *School Science Review*, **57**, *No. 198*, 22–37

6.3 Crellin, J. R. (1978) 'Environmental measurement in schools: electronics in field studies', *Journal of Biological Education*, **12**, 190–8

6.4 Dowdeswell, W. H. and Humby, S. R. (1953) 'A photo-voltaic light meter for school use', *School Science Review*, **35**, *No. 125*, 64–70

6.5 Westlake, D. F. and Dawson, F. H. (1975) 'The construction and long-term field use of inexpensive aerial and aquatic integrating photometers', in *Light as an Ecological Factor*, British Ecological Society, Blackwell

6.6 Bishop, O. N. (1975) 'A photometer for biological investigations', *School Science Review*, **57**, *No. 199*, 241–53

6.7 Dowdeswell, W. H. (1981) *The Life of the Meadow Brown*, Heinemann

6.8 Solomon, M. E. (1945) 'The use of cobalt salts as indicators of humidity and moisture', *Annals of Biology*, **32**, 75–85

6.9 Obtainable from B.D.H. Chemicals Ltd, Poole, Dorset

6.10 Obtainable from E.F.S., 15 Wulfruna Gardens, Wolverhampton WV3 9HZ

6.11 Gill, B. F. (1977) 'A plastic syringe method for measuring dissolved oxygen in the field or laboratory', *School Science Review*, **55**, *No. 204*, 458–60

6.12 Richardson, J. (1981) 'Oxygen meters: some practical considerations', *Journal of Biological Education*, **15**, 107–16

6.13 Crellin, J. R. and Tranter, J. (1978) 'Monitoring the environment: the use of electronic meters and chemical or bacteriological tests', *Journal of Biological Education*, **12**, 291–304

6.14 Lapworth, C. H. (1981) 'Measuring salinity by conductivity', *Journal of Biological Education*, **15**, 186–7

6.15 Lind, O. T. (1974) *Handbook of common methods in limnology*, Mosby, Saint Louis

6.16 Wood-Robinson, C. (1981) 'Surveying Rocky Shores', *Journal of Biological Education*, **15**, 100–1

Appendix A

The Winkler method of estimating dissolved oxygen

Chemical principles

The method depends upon the fact that when a manganese(II) salt is added to water containing oxygen in alkaline conditions a proportion of the manganese(II) is oxidised to manganese(III). Upon acidification in the presence of potassium iodide, iodine is liberated in an amount equivalent to the dissolved oxygen present. The free iodine is then estimated by titration against a standard thiosulphate solution, using starch as an indicator.

The following adaptation of the Winkler method is that of Gill [6.11]. Experience both in the field and the laboratory has shown it to be reliable and easy to use.

Reagents

A – 40 per cent w/v manganese(II) chloride solution

B – 80 g sodium hydroxide and 2.5 g potassium iodide in 250 cm³ water

C – Phosphoric(V) acid – approximately a 50 per cent solution, the exact concentration is not critical

D – 0.025 M (M/40) sodium thiosulphate(VI) solution

E – 0.25 per cent w/v starch in saturated sodium chloride solution.

Sodium thiosulphate(VI) solution can be conveniently prepared from ampoules sold for volumetric analysis. A 0.5 M stock solution prepared from one ampoule will keep for several months if stored in a stopped bottle in a refrigerator.

Apparatus

1 A 20 cm³ plastic syringe in which the reactions will take place. In shallow water it can also serve as the water sampler.

2 A 1 cm³ plastic syringe. This is used as a micro-burette to inject the thiosulphate(VI) into the reaction vessel in the final titration of iodine.

3 A coupling for the two syringes. This is made by removing the plastic fitting from a hypodermic needle and mounting it on a second needle as in Fig. 6.21. The fitting can be removed quite easily by gripping the needle in a vice and pulling the fitting off with a pair of pliers.

Procedure

1 A 14 cm³ water sample is required in the large syringe (the needle is not required). To avoid the inclusion of air bubbles, first fill the whole syringe *slowly*. Now invert it and tap it so that any air bubbles accumulate at the nozzle. Expel any air and excess water until 14 cm³ remain.

2 Add a small amount of reagent A. The amount is not critical but a syringe nozzle full is convenient. Wipe the nozzle.

3 Add a nozzle full of reagent B and allow the liquids to mix. A precipitate of hydroxide will be formed. Wipe the nozzle.

4 Wait for at least a minute while the manganese(II) hydroxide reacts with any oxygen present.

5 Add a nozzle full of reagent C. The precipitate will dissolve, liberating iodine and producing a golden colour. If any precipitate remains, add a little more reagent. At this stage no further reaction with oxygen can take place. To allow space for the addition of liquid in the titration, a few cubic centimetres of air should now be drawn into the syringe.

6 Fill the 1 cm³ syringe (no needle) with reagent D. Invert to remove air bubbles and expel surplus liquid to the 1 cm³ mark. Fit the small syringe to the larger one using the coupling device (Fig. 6.21) with the large syringe uppermost.

7 Inject thiosulphate(VI) solution (reagent D) until the iodine is a pale straw colour.

8 Detach the small syringe and take a few drops of reagent E up into the large syringe. A blue-black colour will develop.

9 Attach the small syringe again and continue injecting thiosulphate(VI) until the blue-black colour just disappears. This is the end point.

10 Note the volume of thiosulphate(VI) which has been added.

The amount of dissolved oxygen is given by:

$$O_2 \text{ concentration (cm}^3 \text{ dm}^{-3} \text{ water)} = \text{volume thiosulphate(VI) (cm}^3) \times 10$$

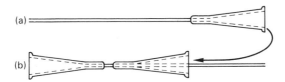

Fig. 6.21 Coupling device for two hypodermic syringes. The plastic fitting is removed from hypodermic needle (a) and pushed onto a second needle (b)

When the dissolved oxygen is low and the amount of iodine liberated at stage 5 is small, stage 7 can be omitted. Under conditions of high oxygen more than 1 cm^3 of thiosulphate(VI) may be required, but this will rarely occur.

For comparative purposes it is useful to express oxygen concentration as a percentage of saturation for a particular temperature. This can easily be done by converting the volume obtained in stage 10 above to STP and referring to Table 6.5 below.

Table 6.5   Oxygen dissolved by distilled water when saturated with air at different temperatures (after Roscoe and Lunt)

Temperature (°C)	*Oxygen at STP* $(cm^3 dm^{-3})$
5	8.68
6	8.49
7	8.31
8	8.13
9	7.95
10	7.77
11	7.60
12	7.44
13	7.28
14	7.12
15	6.96
16	6.82
17	6.68
18	6.54
19	6.40
20	6.28

Using information

As was emphasised in the previous chapter, both how to obtain the relevant information and how to use that information, are important in the planning and implementation of an ecological study. They are separated here merely for the purpose of convenience. All ecological studies involve an element of observation and description, but if we are to identify and analyse a problem with any precision it is necessary to have recourse to quantitative methods. In this chapter we shall consider some of the techniques available for presenting and interpreting ecological data.

Methods of presentation

The presentation of ecological information involves its assembly in a form in which it is most readily comprehensible and available for further analysis. The two methods most widely used are tabulation and graphical presentation.

Tabulation

This consists of the arrangement of figures in columns and by categories, and is the procedure most widely adopted in scientific work. It has the advantages of clarity and ready accessibility but is less effective as a means of showing the relationships between different variables.

Graphical presentation

(*i*) *Graphs*
These provide a ready means of illustrating pictorially the relationship between two or more variables. Thus periodic measurements of abiotic factors such as temperature or light are conveniently recorded in this way. (It is a convention that time is always plotted on the *x* axis.)

When constructing graphs, the question of whether there is justification for joining successive points is often overlooked. In ecology, the data likely to be presented as graphs are usually of two kinds.

(a) The quantity to be measured may be continuous but erratic, for example, light intensity or temperature (Fig. 7.1). It is permissible here to join the points with straight lines, thus implying continuity of the variable but ignorance of its intervening fluctuations.

(b) Many ecological problems are concerned with relationships between populations and one or more factors, either biotic or abiotic. As we have seen already, the interactions between organisms and their environment can be exceedingly complex and it is therefore unlikely that the effect of one factor will be so powerful as to mask all the others. The type of graph obtained in such circumstances is not the straight line or uniform curve characteristic of

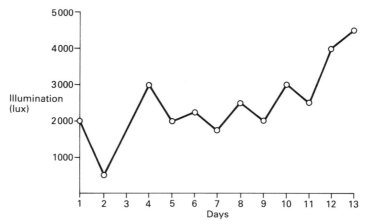

Fig. 7.1   Illumination at the surface of a pond on successive days (at 10.00 hours) recorded as a graph

mathematical functions but rather the sort of diagram shown in Fig. 7.2. Here, there is clearly a general relationship between current speed and density of population. We are thus justified in drawing as nearly as possible an ideal line between the scatter of points, indicating an association between the variables observed. The possible nature of this association (correlation) will be discussed later in the chapter (p. 136).

(ii) Bar-charts

When studying the characteristics and distribution of populations and communities, it is often necessary to handle grouped data such as the density of successive samples taken from an animal population over a period of time, or the frequency of a plant species at different points along a belt transect. Such information can be conveniently represented by a bar-chart as in Fig. 7.3, which shows the distribution of small planktonic crustaceans in successive samples from

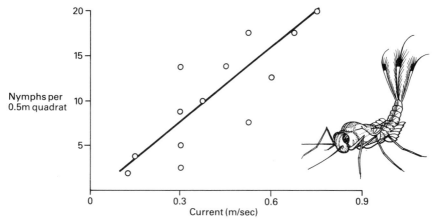

Fig. 7.2   Relationship between speed of current and distribution of mayfly nymphs in a stream, illustrating a possible correlation

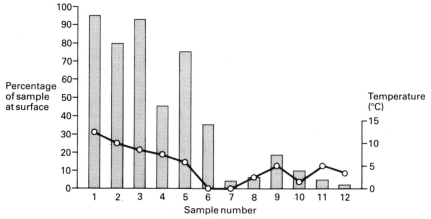

Fig. 7.3 Relationship between plankton distribution and temperature in a small pond, expressed as a bar-chart

a pond. It is often convenient to superimpose a time graph on a bar-chart if the two quantities being studied are thought to be related, as in Fig. 7.3.

(iii) Histograms

While the bar-chart is used to record discontinuous data, the histogram provides a means of recording continuity, as in the analysis of a sample in terms of some variable such as height, weight or colour (see Fig. 7.4). For simplicity of presentation the data are grouped into categories. In the study of groundsel height (Fig. 7.4) these were at 20 mm intervals, the smallest being 0 to 20 mm, the next 20+ to 40 mm, and so on. The histogram is a graphical method of recording actual numbers (as opposed to percentages which should be shown as dots joined by straight lines) and is a convenient way of determining at a glance the structure of a

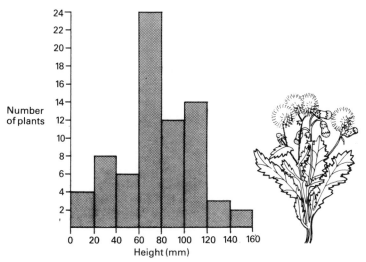

Fig. 7.4 Histogram recording frequency distribution of height in a 1 m² sample of a population of groundsel, *Senecio vulgaris*

population in respect to a particular characteristic. Thus as a population of groundsel plants gets older, the proportion of tall individuals will tend to increase and the peak of the figure will shift from left to right. By now the oldest plants may well have seeded, giving rise to a second peak among the small individuals as germination proceeds. Again, animal populations which are subject to heavy predation by birds, such as those of snails, sometimes exhibit curious gaps in the frequency distribution of shell size. This may well prove to be evidence of preferential selection by predators and serves to illustrate the important point that histograms are not only methods of analysis but may also be valuable in highlighting the existence of ecological problems that could repay further study.

Comparing the relative merits of tabulation and graphical methods of presentation, we can conclude that while tables provide the most efficient way of presenting data for further analysis, graphs, bar-charts, and histograms are preferable as means of revealing patterns and relationships. Moreover, their value lies not only in the ordering of ecological data but in the contribution they can sometimes make both to the identification of new problems and to a clearer understanding of those already under consideration.

The mean

In ecological studies, one of the commonest needs is a way of comparing samples either of different populations of the same species, or of the same population in varying circumstances. One obvious approach is to determine the mathematical average or **mean** of each sample. However, although two means may appear somewhat similar, the figures comprising them can fluctuate considerably, as the following example shows. Two adjacent populations of the white-lipped snail, *Cepaea hortensis*, were sampled, with the intention of comparing the size distribution of their shells. The hypothesis being tested was that since one population (A) occupied a more sheltered location than the other (B) its pattern of reproduction might be different. Two samples, each of approximately 200 animals, were collected and the results were as follows:

Population A		Population B	
Diameter of shell (mm)	Number of snails	Diameter of shell (mm)	Number of snails
11	7	11	2
12	16	12	8
13	24	13	6
14	11	14	10
15	42	15	38
16	38	16	52
17	29	17	40
18	20	18	23
19	12	19	25
20	4	20	6
21	1	21	3
	Total 204		Total 213

$$\text{Mean} = \frac{\text{Total snails}}{\text{Number of classes}} = \frac{204}{11} = 18.55 \qquad \text{Mean} = \frac{213}{11} = 19.36$$

The means of the two populations appear to differ, suggesting that the sheltered one (A) contained a higher population of young individuals than the other. However, it could be argued that with the size of the two populations amounting to many thousands, a sample of only 200 was hardly representative and may therefore have been subject to appreciable sampling error. In order to make our comparison as valid as possible, two factors need to be taken into account:

(i) the number of observations made (snails examined)
(ii) the range of variation within each group.

There is no easy method of deciding the appropriate size of a sample. Ideally, in order to avoid sampling error, the whole population should be collected but this is obviously impossible. Time and the facilities available are usually crucial factors influencing such a decision. However, valuable help can also be obtained by referring to various quantitative techniques.

Sampling error (standard deviation)

If we obtained a sample of continuously variable organisms such as snails and plotted the frequency distribution of their shell size, we would find that only a few lay on the arithmetic mean, the remainder being distributed on either side of it. The result would be a bell-shaped curve such as that in Fig. 7.5. This is known as a curve of **normal distribution** and it has certain characteristic features. It is symmetrical about the mean; the area beneath any portion of the curve is proportional to the number of observations associated with that part; irrespective of the scale on which it is drawn and the units used, the curve has certain fixed mathematical properties.

The extent to which this distribution is characteristic of a population as a whole will depend on the proportion sampled and the randomness of the sampling technique. A convenient mathematical model for assessing error is that shown in Fig. 7.5, where the whole population is represented by the area under the curve (A). The small shaded portion to the left (a) represents part of the population (in our example the smallest shells). Thus a/A will represent the likelihood that an individual picked at random from the population will belong to the range of shells belonging to this small group. In order to quantify the deviation of a sample from the mean ($\bar{x}$) we need to calculate a statistical quantity known as the **standard deviation (σ)**. As can be seen from the graph (Fig. 7.5), roughly 68 per cent of the

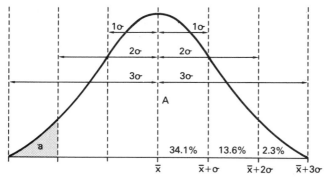

Fig. 7.5 Standard deviation and the normal curve. $\bar{x}$ = mean; σ = standard deviation

population falls within one standard deviation from the mean, 95 per cent within two standard deviations, and 100 per cent within three standard deviations.

The standard deviation is given by the formula:

$$\sigma = \sqrt{\frac{\sum (x - \bar{x})^2}{N}}$$

where x = number in size class
$\bar{x}$ = mean
N = number of classes
$\sum$ = sum

We can now carry our analysis of the *Cepaea* data a step further.

Classes (mm)	x	$x - \bar{x}$	$(x - \bar{x})^2$
Population A			
11	7	−11.6	134.6
12	16	−2.6	6.8
13	24	5.4	29.2
14	11	−7.6	57.8
15	42	23.4	547.6
16	38	19.4	376.4
17	29	10.4	108.2
18	20	1.4	2.0
19	12	−6.6	43.6
20	4	−14.6	213.2
21	1	−17.6	309.8
	204		1829.2

$\sum x = 204$
Mean $(\bar{x}) = 18.6$
$\sum (x - \bar{x})^2 = 1829.2$
$$\sigma = \sqrt{\frac{1829.2}{11}} = 12.9$$

Thus, the true mean lies within the range 18.6 ± 12.9.

Similar treatment of the samples from population B gives $\bar{x} = 19.4$ and $\sigma = 16.6$. So the true mean lies within the range 19.4 ± 16.6.

Standard error of the mean

The results quoted above are typical of ecological investigations where successive samples from the same or different populations vary in their means. This is to be expected. We can make an estimate of the range of the deviation from the mean by calculating a statistic known as the **standard error** (*SE*). This is given by the formula:

$$SE = \frac{\sigma}{\sqrt{\sum x}}$$

where σ = standard deviation
$\sum x$ = total of the sample

For population A, $SE = \dfrac{12.9}{\sqrt{204}} = 0.9$

Referring to Fig. 7.5, we see that approximately 95 per cent of the population falls within the range:

$$\bar{x} \pm 2 \times SE = 18.6 \pm 1.8$$

This is known as a 95 per cent confidence limit. Put another way, it means that for 95 per cent of all similar samples the mean will lie within the range 16.8–20.4, or 5 per cent (i.e. 1 in 20) might be expected to fall outside this range. For population B, $SE = 1.1$.

Standard error and the comparison of two means

If we wish to estimate the difference between the means of two populations where the standard deviations of the samples are known, we must calculate the standard error of the difference, which is given by:

$$SE \text{ of difference} = \sqrt{\dfrac{\sigma_1^{\,2}}{\sum x_1} + \dfrac{\sigma_2^{\,2}}{\sum x_2}}$$

where σ_1 and σ_2 are the standard deviations
$\sum x_1$ and $\sum x_2$ are the total of each sample.

In our example,

$$SE \text{ of difference} = \sqrt{\dfrac{12.9^2}{204} + \dfrac{16.6^2}{213}} = 2.1$$

Thus, at the 95 per cent confidence limit the difference between the two means $(19.4 - 18.6 = 0.8)$ lies within the range $0.8 \pm 2 \times 2.1 = 0.8 \pm 4.2$.

From what has been said so far it will be seen that when comparing the differences between samples there is no such thing as a certain result. Certainty could only be achieved by examining the whole population, which is generally impossible. The best we can usually achieve is an estimate of the chance that a particular situation will occur. This leads us to the idea of **probability** – the likelihood that a given result will, by chance, be equalled or surpassed. As was mentioned earlier in connection with Fig. 7.5, the likelihood of an individual selected at random from the total population A belonging to the small group a will be a/A. This, then, provides us with an estimate of the probability of the occurrence taking place.

Reference to statistical tables shows that the approximate probability of an observed deviation exceeding different multiples of the standard error is as follows:

exceeding $SE \times 1$ by chance is $1/3$
exceeding $SE \times 2$ by chance is $1/22$
exceeding $SE \times 3$ by chance is $1/370$
exceeding $SE \times 4$ by chance is $1/17\,000$.

A difference between two values is judged to have attained **statistical significance** if it equals or exceeds twice the standard error. That is to say, the odds are roughly 20:1 against the result being due to chance, so that, on average, once in 20 times we would expect the observed difference to be less than $2 \times SE$. The reason for selecting this value as the point at which a difference begins to attain significance is explained by the table above, for it stands at the point where the curve of probability starts to become much steeper, so that only a slight excess over it means that the chance of obtaining the result by luck is quite small.

Returning once more to our *Cepaea* example, the observed difference between the two means was 0.8, and $2 \times SE = 4.2$. There is thus no evidence of a difference between the two populations, and our original hypothesis regarding different rates of reproduction is not upheld. When making comparisons of this kind, the question often arises as to the number of samples to be taken. Reference to the standard error can often help in providing an answer. In the present example, the magnitude of the divergence between the observed difference and $2 \times SE$ suggests a fairly conclusive result. However, if time and circumstances had permitted, it would have been worth taking a few more samples to see if the departure was maintained. Had the divergence been much less (say the observed difference was about 4 instead of 0.8) and the conclusion therefore equivocal, more samples would have been needed. An indispensable item of equipment for the field ecologist is, therefore, a pocket calculator.

The problem of small samples

The previous calculations of standard deviation and standard error depended upon an important assumption, namely that the distribution of individuals conformed to a curve of normal distribution (i.e. was at random) or nearly so. This was justified provided the size of the samples was reasonably large, about 30 or more. The smaller a population becomes, the more will its distribution tend to deviate from normality. The extent of this deviation can be estimated by means of the so-called *t*-**distribution**. This enables us to determine the scatter within a population and to compare different populations in precisely the same way as the standard deviation and standard error. Since a common requirement in ecology is to compare populations from which a series of samples have been obtained, a typical example has been selected to illustrate the procedure involved. The *t* test involves a rather tedious calculation, but this can be carried out quite quickly using a pocket calculator. It can be applied equally well to large samples as to small ones, but as the numbers increase the *t*-distribution approaches a normal distribution more and more closely so that the extra effort is hardly worthwhile.

The formula is expressed as:

$$t = \frac{(\bar{x}_1 - \bar{x}_2)\sqrt{\dfrac{1}{N^1}}}{\sqrt{\sum (x_1{}^2) - \dfrac{(\sum x_1)^2}{N_1} + \sum (x_2{}^2) - \dfrac{(\sum x_2)^2}{N_2}}}$$

and

$$\sqrt{\frac{1}{N_1}} = \sqrt{\frac{N_1 N_2 (N_1 + N_2 - 2)}{N_1 + N_2}}$$

where $\sum x_1$ and $\sum x_2$ are the sums of the two series of samples
  $\bar{x}_1$ and $\bar{x}_2$ are their means
  N_1 and N_2 are the number of samples in each group

The following problem is typical of the sort requiring this kind of statistical treatment. A comparative study was made of earthworm populations in two neighbouring plots of ground, one of which had been manured while the other had not. Sampling was carried out using one metre quadrats and applying a standard quantity of dilute formaldehyde solution which caused the worms to come to the surface where they were collected and counted.

Plot 1 (with manure)		Plot 2 (no manure)	
Sample	Number of earthworms	Sample	Number of earthworms
1	5	1	4
2	9	2	3
3	12	3	6
4	9	4	8
5	10	5	5
6	7	6	3
7	5	7	4
8	8	8	5
9	4		
	Total 69		Total 38
Mean $(\bar{x}_1)$ = 7.67		Mean $(\bar{x}_2)$ = 4.75	

The hypothesis to be tested is that the plot with manure supported a higher density of earthworms than that without.

For Plot 1

$$N_1 = 9$$
$$\sum (x_1) = 69$$
$$\sum (x_1)^2 = 585$$
$$\bar{x}_1 = 7.67$$

For Plot 2

$$N_2 = 8$$
$$\sum (x_2) = 38$$
$$\sum (x_2)^2 = 200$$
$$\bar{x}_2 = 4.75$$

For the two series of samples $t = 2.68$

The next step is to determine the number of **degrees of freedom** (N). The concept depends on the assumption that each difference between x and $\bar{x}$ is independent. But because $\sum (x - \bar{x}) = 0$, only $N - 1$ values of x can vary independently; the last (Nth) value of x must be determined by the remaining values. Put another way, the number of degrees of freedom is the least number of independent variables which must be given values before the state of a system can be completely determined. The figure can be calculated from the formula:

$$N = N_1 + N_2 - 2$$

In our example this gives a value of 15. Reference must now be made to the table of t (see Appendix A at the end of this chapter, p. 146) showing the values for different numbers of degrees of freedom corresponding to a probability p. For $N = 15$ and a value of $t = 2.60$, $p = 0.02$ (i.e. 1 in 50). The calculated estimate (2.68) exceeds this, being nearer 2.60 than 2.95 (see table). Hence the data available support our hypothesis that the earthworm densities in the two plots were different. As in the previous calculations using large samples, estimated values of t will sometimes hover on the brink of significance, perhaps being equal to or near 0.05. Equivocal results of this kind need to be interpreted with caution and indicate that further samples will be required before reliable conclusions can be drawn.

Comparing more than two means

So far, we have been concerned with the comparison of only two means, but suppose we need to compare more than two. One way would be to select the means in pairs and calculate the standard error for each combination. Alternatively, if the sample was small, we could perform the same operation using the *t* test. But this would be a tedious and cumbersome undertaking. To get over the problem, a different kind of test called an **analysis of variance** is used. This takes account of the total variation (variance) of the sets of samples. The total variance consists of the sum of two components, the variance occurring *between* series of samples and that existing *within* them. Details of the procedures involved in the calculation need not concern us here but can be obtained from books on statistics such as those included in the Bibliography (p. 295) under the heading 'Ecological investigation'.

Relationship between variables: correlation

When studying the spatial distribution of a species like the freshwater shrimp, *Gammarus pulex*, in a stream, it sometimes happens that a single variable such as current speed appears to be exerting a predominant influence. The question then is whether a relationship (**correlation**) exists between that factor and animal density. An approximate estimate of the degree of correlation between two variables can be obtained by constructing a scatter diagram (Fig. 7.6). If the relationship is perfect, as in certain mathematical functions, all the points will fall on a straight line (a). In ecology, the more usual situation is likely to be represented by b–d where the two variables are related positively, negatively or not at all. However, circumstances will often arise when it is not easy to decide whether the scatter of points justifies the drawing of a straight line between them. Moreover, following our previous practice, there is much to be said for reducing relationships or the lack of them to more precise and quantitative terms. For this

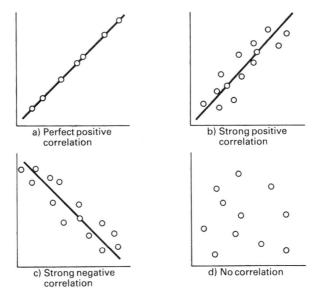

Fig. 7.6 The use of scatter diagrams to show the degree of correlation between two variables

purpose we use a statistic known as the **correlation coefficient** (r); the greater the value of r, the closer the degree of relationship. Thus for a perfect linear correlation (Fig. 7.6, a), $r = 1$. A typical example will illustrate the procedure involved.

A class was studying the relationship between current speed in a stream and the distribution of the crustacean, *Gammarus pulex*, in order to test the hypothesis that the two were positively related. A typical set of results was as follows:

Sampling site	Current speed (cm s^{-1})	Number of Gammarus in standard sample
1	10	4
2	30	6
3	40	15
4	60	5
5	65	24
6	80	12
7	90	32
8	95	27

To calculate r, the necessary equation is:

$$r = \frac{\sum ((x - \bar{x})(y - \bar{y}))}{\sqrt{(\sum (x - \bar{x})^2 . \sum (y - \bar{y})^2)}}$$

where x and y are the two variables
$\bar{x}$ and $\bar{y}$ are the means of the two sets of variables

This rather cumbersome notation can be simplified if for $(x - \bar{x})$ we write dx (d denoting the deviation between the two), and for $(y - \bar{y})$ we substitute dy. The equation then becomes:

$$r = \frac{\sum dx . dy}{\sqrt{d^2x . \sum d^2y}}$$

The data can now be rearranged for greater convenience as follows:

1 Current speed (cm s^{-1}) (x)	2 Number of Gammarus (y)	3 Deviation of x from mean (dx)	4 Deviation of y from mean (dy)	5 (d^2x)	6 (d^2y)	7 ($dx . dy$)
10	4	48.75	11.63	2376.56	135.26	566.96
30	6	28.75	9.63	826.56	92.74	276.86
40	15	18.75	0.63	351.56	0.40	11.81
60	5	1.25	10.63	1.56	113.00	13.29
65	24	6.25	8.37	39.06	70.06	52.31
80	12	21.25	3.63	451.56	13.18	77.14
90	32	31.25	16.37	976.56	267.98	511.56
95	27	36.25	11.37	1314.06	129.28	412.16
$\sum x = 470$ $\bar{x} = 58.75$	$\sum y = 125$ $\bar{y} = 15.63$			$\sum d^2x =$ 6337.48	$\sum d^2y =$ 821.90	$\sum dx . dy =$ 1922.09

Note that in calculating the values for columns 3 and 4, minus signs are omitted.

Thus

$$r = \frac{1922.09}{\sqrt{6337.48 \times 821.90}} = 0.84$$

In order to determine the level of probability corresponding to this value of r we must refer to the table of correlation coefficients (Appendix B, p. 147). However, it is first necessary to decide on the number of degrees of freedom, which is given by $N - 1$ when N is the number of observations. Here the number of degrees of freedom is $8 - 1 = 7$. It will be seen that for a value of $r = 0.84$, with 7 degrees of freedom, p (probability) lies between 0.01 and 0.001. Thus, there is strong evidence in support of the initial hypothesis, since the correlation between current and *Gammarus* population density is in excess of the 99 per cent confidence level.

The chi-squared (X^2) distribution

Earlier, we saw how statistics such as the standard deviation and standard error can be useful in comparing two variables (i.e. where there is one degree of freedom). Their use rests on the assumption that the distribution of each variable is normal or nearly so. But it will frequently happen in ecological studies that we know nothing about the distribution of a group of organisms, the only data available being the number of individuals falling into particular categories, such as the size-groups of plants and animals belonging to different populations. In such circumstances, the procedures already discussed are often inappropriate and it is necessary to make use of a statistic called **chi-squared** (χ^2). This calculation provides one of the most useful tests of statistical significance in biology. It can be used to compare the distribution of individuals within separate categories either with some theoretical value or with different parts of a population.

The mathematical derivation of χ^2 is complicated and will not be discussed further here. For the relevant information the appropriate books on statistics should be consulted (see Bibliography). The basis of the procedure is as follows. If O is the observed frequency of any one variable and E the frequency that would be expected as a result of some hypothesis, then χ^2 is obtained by dividing the square of the difference between O and E by E and summing the quotients for each category into which the variable falls. Thus,

$$\chi^2 = \Sigma \left[\frac{(O - E)^2}{E} \right]$$

Sometimes there are grounds for expecting a particular result, for instance when testing for a genetic ratio or for the equality of sexes. But in ecology such instances are comparatively rare and there are seldom grounds for predicting that a particular relationship between two variables will be achieved. One of the features of χ^2 calculations is that in such circumstances it is possible to generate one's own expectation based on the information available.

The essential point about χ^2 is that it involves a comparison of observed values with those expected as a result of some theory. The hypothesis to be tested is that any differences could be accounted for by chance. For any biological event it is usually unlikely that observation will accord exactly with expectation. However, there must come a point at which chance can no longer account for differences between the two and it is this point which χ^2 assists us in identifying. Since our initial hypothesis is that no difference exists that cannot be ascribed to chance, it is known as a **null hypothesis**. Clearly, if we can disprove a null hypothesis, it follows that the variables must be associated with one another. A few typical examples will illustrate how χ^2 can be used in different ecological contexts.

Two variables with an expected ratio

Situations of this kind are relatively rare in ecology, being mainly concerned with differences under the control of clear-cut genetic mechanisms such as sex ratios and characteristics exhibiting genetic polymorphism (p. 9). For instance, the primrose, *Primula vulgaris*, possesses three different kinds of flower, pin, thrum, and long homostyle (Fig. 7.7 and Fig. 1.2). Pin and thrum are controlled by a single pair of alleles (*s* and *S*), thrum being dominant. In nature, all thrum plants are heterogygous (*Ss*), so normal pollination results in a back-cross, 50 per cent of each form being produced. There is thus some justification for expecting equality between the two forms in wild populations. A count of primroses flowering on the edge of an oak wood in Hampshire produced the following result:

Number of plants	Pin	Thrum
254	138	116

What evidence is there of a departure from equality? Our null hypothesis is that there is no significant difference between the proportions of pin and thrum. The calculation is conveniently presented in the form of a table:

	Pin	Thrum	Totals
Observed (*O*)	138	116	254
Expected (*E*)	127	127	254
O − *E*	11	− 11	
(*O* − *E*)²	121	121	

$$\chi^2 = \Sigma \left[\frac{(O - E)^2}{E} \right] = 0.94 + 0.94 = 1.88$$

The number of degrees of freedom (*N*) = 1 (see p. 135). This can be written as $\chi^2_{(1)}$. Turning to the table of χ^2 (Appendix C, p. 148) we see that for $N = 1$ and $\chi^2 = 1.88$, *p* (probability) lies between 0.20 and 0.10. From the data available, therefore,

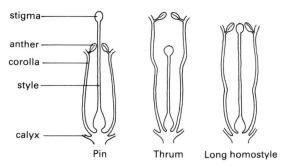

Fig. 7.7 Flower structure in the primrose, *Primula vulgaris*

there is no evidence of a divergence from equality (this would be indicated if $p = 0.05$ or less). Bearing in mind the size of the sample it is unlikely that the addition of further samples would have changed the result.

With only one degree of freedom, the primrose data could equally well have been analysed using the standard error (p. 132), but the advantage of χ^2 is that it is more sensitive and little more laborious to calculate.

Two variables, with no expected ratio

A common situation in ecology is when two variables are being studied each of which can be divided into two classes. The data for primrose populations once again provides a typical example. In a nearby population to the one already considered (p. 139) a count of primroses on the same day consisted of,

Total	Pin	Thrum
550	161	389

The null hypothesis to be tested is that there is no difference between the two populations, any divergence between them being due to chance. Arranging the data to be compared as shown below gives us a table of **general contingency**:

	Pin	Thrum	Totals
Population 1	138	116	254
Population 2	161	389	550
	299	505	804

The two variables (pin and thrum) are each divisible into two categories only (populations 1 and 2), hence we speak of a 2 × 2 table. The totals of the vertical and horizontal columns should add up to the same grand total (804).

Theoretically, the next step should be to deduce the expected values since we have no justification in assuming that any relationship exists between the respective variables. But in a 2 × 2 table, a simplified method of calculation avoids this necessity. This is best explained by re-writing the table using letters.

	x_1	x_2	Totals
y_1	a	b	$\sum y_1$
y_2	c	d	$\sum y_2$
	$\sum x_1$	$\sum x_2$	T

$$\chi^2 = \frac{(ad - bc)^2 T}{(\sum x_1)(\sum x_2)(\sum y_1)(\sum y_2)}$$

In our example,

$$\chi^2 = \frac{[(138 \times 389) - (116 \times 161)]^2\ 804}{299 \times 505 \times 254 \times 550} = 4.67$$

We now need to determine the number of degrees of freedom available. This is given by the formula $(c - 1)(r - 1)$, where c and r are columns and rows respectively. It follows that the number of degrees of freedom (N) in any 2 × 2 table must always be 1. From the table of χ^2 (Appendix C) a value of $\chi^2_{(1)} = 4.67$ corresponds to a probability (p) of between 0.05 and 0.02. So there is evidence to disprove the null hypothesis, which indicates a significant difference between the two populations in respect of the incidence of pin and thrum plants. An important point to note here is that statistical treatment of this kind is concerned with estimates of probability only. It gives no indication of *why* differences occur, only that they exist. Such information is valuable not so much in the solution of ecological problems as in highlighting their existence and the need for further study.

More than two variables

In ecological studies, circumstances will frequently arise where two or more variables are being studied, each of which is divisible into more than two classes. Rarely, however, will there be grounds for expecting a particular ratio.

Returning to the data on primroses, mention was made earlier (p. 139) of a third form of flower, long homostyle. This form provides a mechanism for self-pollination and therefore inbreeding. Turrill [7.1] has recorded counts of populations in Somerset where the three types grew together, two of his samples being:

Population	Pin	Thrum	Long homostyle
1	102	11	210
2	145	15	468

Rearranging the data for comparison as before we get:

	Pin	Thrum	Long homostyle	Totals
Population 1	102	11	210	323
Population 2	145	15	468	628
	247	26	678	951

This is known as a $2 \times n$ table and the method of calculating χ^2 from it can best be explained by substituting letters as before.

	x_1	x_2	x_3	Totals
y_1	a_1	a_2	a_3	Σy_1
y_2	b_1	b_2	b_3	Σy_2
	Σx_1	Σx_2	Σx_3	T

Three values of χ^2 are now calculated. For x_1 (pin) the equation is:

$$\chi^2 = \frac{[(\Sigma y_2 \times a_1) - (\Sigma y_1 \times b_1)]^2}{(\Sigma x_1)(\Sigma y_1)(\Sigma y_2)}$$

and similar equations can be deduced for x_2 (thrum) and x_3 (long homostyle).

The sum of these expressions will give the total value of χ^2. Thus,

$$\frac{[(628 \times 102) - (323 \times 145)]^2}{247 \times 323 \times 628} = 0.59$$

$$\frac{[(628 \times 11) - (323 \times 15)]^2}{26 \times 323 \times 628} = 0.81$$

$$\frac{[(628 \times 210) - (323 \times 468)]^2}{678 \times 323 \times 628} = 2.71$$

$$\chi^2 = 4.11$$

The number of degrees of freedom (N) is calculated from $(c-1)(r-1)$, which gives 2 (see p. 140). Turning to the table of χ^2 (Appendix C), it will be seen that for $N = 2$, a χ^2 value of 4.11 lies between p 0.2 and 0.1. There is therefore no evidence of a difference between the two populations and the null hypothesis is supported.

Should we wish to compare the incidence of the three flower types in more than two colonies (Turrill compared five) we would need to construct an $n \times n$ table in the same way as before. However, there is no quick way of calculating χ^2; the value for each cell must be determined separately and all then added together. The first step is to estimate the expected value for each observed frequency. Referring to the previous table (p. 141), that corresponding to a_1 is given by

$$\frac{(\sum x_1)(\sum y_1)}{T}$$

Similarly, that for b_2 is

$$\frac{(\sum x_2)(\sum y_2)}{T}$$

The figures so obtained are inserted in each cell (in brackets to distinguish them from the observed values) and χ^2 is calculated from the formula:

$$\frac{(O - E)^2}{E} \quad \text{(see p. 138).}$$

As before, the number of degrees of freedom (N) will be obtained from $(c-1)(r-1)$.

Although somewhat tedious to calculate, an $n \times n$ table can be invaluable in ecological work, not only as a means of testing the relationship between two comparable sets of variables but also for determining whether a series of data is homogeneous or heterogeneous. We thus have a means of answering the question which is often so difficult to resolve by inspection alone, namely, whether we are justified in adding the results of a number of related samples which appear superficially to resemble one another? We must remember that it is committing statistical heresy to lump together data that are heterogeneous!

Computers in ecology

The rapid increase in the availability of computers and their accompanying software has enabled ecological situations to be simulated and assessed in a way hitherto virtually impossible without an inordinate expenditure of time and effort. The kinds of calculation which are useful fall into two groups:

(i) lengthy statistical and other arithmetical computations which can be greatly speeded-up
(ii) simulations involving an element of prediction.

Statistical calculations

Here, the computer is used merely as a glorified form of calculating machine. As with all such machines, the reliability of the results depends entirely on the level of the information fed in. As experience has so frequently shown, it is all too easy to make the occasional slip in feeding in large quantities of data. Moreover, as the operator tires, the number of errors can increase alarmingly.

Such considerations serve to underline two important points:

(i) when computing a statistical calculation it is essential to form a rough estimate of the likely answer. An error, say by a factor of 10, can then be instantly detected.
(ii) the sheer speed with which a result is obtained can serve to obscure its meaning. When computing a statistic it is important not only to comprehend its meaning but also to be aware of the limitations of its use in a particular situation.

Computer programs can be invaluable for carrying out the more tedious statistical calculations such as χ^2, $2 \times n$ and $n \times n$ tables (p. 141). Programs are now also available for simpler purposes such as calculation of the standard deviation (p. 131) and correlation coefficient (p. 136), although these can easily be worked out using a pocket calculator.

Estimation of animal numbers
Mention was made earlier (p. 94) of a standard method of estimating the numbers in an animal population using the procedure of mark, release, and recapture. Applied over a period of only a few days the method poses no problems of computation, but in order to follow the fortunes of a population over a longer time, we need to have information on increases, survival, and death-rates at different stages. The problem is one of population dynamics, as discussed below, and numerous computer programs now exist in research which perform the necessary calculations. It is surprising that they have not become more widespread for use at an elementary level.

Simulation and prediction

Some of the most useful (and the most misleading) computer programs in ecology relate to the field of dynamics and the influence of external and internal factors on populations and communities. In effect, we are asking the question, 'What will happen if?'. There are two essential requirements for the success of such programs:

(i) the mathematical models on which they are based must be simple and readily comprehensible;

(ii) they must be related to real situations and represent processes observable in the field. Failure to meet this criterion can result in confusion due to emphasis being placed on mathematical rather than ecological considerations.

Some typical examples of population studies where elementary computer programs already exist are as follows:

(a) *Population growth*
The natural growth of a population follows the pattern of an S-shaped curve. Which factors influence this pattern? The effects of both density-dependent and density-independent factors can be compared and assessed.

(b) *Predator–prey relationships*
The numerical relationships between predator and prey, and between parasite and host represent dynamic systems involving feedback. Thus high predation will reduce the numbers of prey which, in turn, affects the density of predators (p. 49). A computer program can provide tentative answers to such questions as, 'What are the maximum permissible fluctuations in predator and prey density that a community can survive?'.

(c) *Population dynamics*
The size of populations in a habitat such as a pond depends on the interactions of a number of factors both abiotic and biotic. Among the latter, the most important is food. What are the effects of variations in the availability of different parts of a food chain?

(d) *Competition*
Plant species compete with one another both inter- and intraspecifically. What are the relative effects of different aspects of these two kinds of competition on the nature of a community?

(e) *Management problems*
In stocking a reservoir with trout, what can be regarded as an optimum density of population and how should it be maintained? Several factors are involved here, notably the amount of fishing permitted and the productivity of the reservoir. There is also the question of the rate of any restocking to allow for natural wastage. Such problems provide excellent examples of the applied side of ecology in action.

(f) *Ecological genetics*
The nature of the gene pool of a population is of interest to ecologists since it represents a fund of variation that is constantly being modified by the non-random survival of individuals due to natural selection. In the past, simple models have been constructed using coloured beads, but these are limited in their scope and tedious to operate. Computer programs enable rapid studies to be made of the effects of selection operating at different levels of intensity, also of genetic drift in both large and small populations.

In conclusion, it is perhaps worth stressing once again that the quality of the information obtainable from a computer program is a function of the quality of the data fed into it. The previous account has, I hope, served to highlight some of the areas in ecology where such quality has already been achieved.

Summary

1 The two most commonly used methods of presenting ecological data are tabulation and graphical analysis. Graphical presentation includes graphs, bar-charts, and histograms.

2 One of the simplest ways of comparing sets of samples is by their mathematical average or mean.

3 The standard deviation represents the scatter of variation about the mean of a sample and assumes a normal distribution. It is also a method of determining error due to sampling.

4 The standard error provides an estimate of the range of deviation in a sample from the mean. It can be used to compare the means of two or more samples and to indicate the probability that an observed difference could be due to change.

5 The concept of statistical significance depends upon the probability of a particular result being achieved. A difference is usually said to be statistically significant if the odds against it occurring by chance equal or exceed 1 in 20 (0.05).

6 Comparing small samples presents particular problems since variation tends to deviate from a normal distribution. The *t*-distribution provides a useful method of overcoming this difficulty.

7 The relationship between two variables (correlation) can be gauged approximately by graphical methods. A more precise and quantitative estimate can be obtained by calculating the correlation coefficient.

8 Chi-squared (χ^2) provides a valuable means of comparing two variables when the nature of their distribution is uncertain (i.e. whether it is normal or not). This is usually the situation in ecological studies, where the statistic has many applications.

References

7.1 Turrill, W. B. (1948) *British Plant Life*, New Naturalist, Collins

Appendix A　Table of *t*

					Probability, *p*						
N	0.90	0.80	0.70	0.50	0.30	0.20	0.10	0.05	0.02	0.01	0.001
1	0.16	0.33	0.51	1.00	1.96	3.08	6.31	12.71	31.82	63.66	636.62
2	0.14	0.29	0.45	0.82	1.39	1.89	2.92	4.30	6.97	9.93	31.60
3	0.14	0.28	0.42	0.77	1.25	1.64	2.35	3.18	4.54	5.84	12.94
4	0.13	0.27	0.41	0.74	1.19	1.53	2.13	2.78	3.75	4.60	8.61
5	0.13	0.27	0.41	0.73	1.16	1.48	2.02	2.57	3.37	4.03	6.86
6	0.13	0.27	0.40	0.72	1.13	1.44	1.94	2.45	3.14	3.71	5.96
7	0.13	0.26	0.40	0.71	1.12	1.42	1.90	2.37	3.00	3.50	5.41
8	0.13	0.26	0.40	0.71	1.11	1.40	1.86	2.31	2.90	3.36	5.04
9	0.13	0.26	0.40	0.70	1.10	1.38	1.83	2.26	2.82	3.25	4.78
10	0.13	0.26	0.40	0.70	1.09	1.37	1.81	2.23	2.76	3.17	4.59
11	0.13	0.26	0.40	0.70	1.09	1.36	1.80	2.20	2.72	3.11	4.44
12	0.13	0.26	0.40	0.70	1.08	1.36	1.78	2.18	2.68	3.06	4.32
13	0.13	0.26	0.39	0.69	1.08	1.35	1.77	2.16	2.65	3.01	4.22
14	0.13	0.26	0.39	0.69	1.08	1.35	1.76	2.15	2.62	2.98	4.14
15	0.13	0.26	0.39	0.69	1.07	1.34	1.75	2.13	2.60	2.95	4.07
16	0.13	0.26	0.39	0.69	1.07	1.34	1.75	2.12	2.58	2.92	4.02
17	0.13	0.26	0.39	0.69	1.07	1.33	1.74	2.11	2.57	2.90	3.97
18	0.13	0.26	0.39	0.69	1.07	1.33	1.73	2.10	2.55	2.88	3.92
19	0.13	0.26	0.39	0.69	1.07	1.33	1.73	2.09	2.54	2.86	3.88
20	0.13	0.26	0.39	0.69	1.06	1.33	1.73	2.09	2.53	2.85	3.85
22	0.13	0.26	0.39	0.69	1.06	1.32	1.72	2.07	2.51	2.82	3.79
24	0.13	0.26	0.39	0.69	1.06	1.32	1.71	2.06	2.49	2.80	3.75
26	0.13	0.26	0.39	0.68	1.06	1.32	1.71	2.06	2.48	2.78	3.71
28	0.13	0.26	0.39	0.68	1.06	1.31	1.70	2.05	2.47	2.76	3.67
30	0.13	0.26	0.39	0.68	1.06	1.31	1.70	2.04	2.46	2.75	3.65

When *N* is greater than 30, *t* may be treated as a normal deviate without serious inaccuracy resulting.

Abridged from *Statistical Tables for Biological, Agricultural and Medical Research* by R. A. Fisher and F. Yates, by permission of the Longman Group Limited, Harlow.

Appendix B   The correlation coefficient, *r*

N	Probability, *p*				
	0.1	0.05	0.02	0.01	0.001
1	0.988	0.997	1.000	1.000	1.000
2	0.900	0.950	0.980	0.990	0.999
3	0.805	0.878	0.934	0.959	0.991
4	0.729	0.811	0.882	0.917	0.974
5	0.669	0.755	0.833	0.875	0.951
6	0.622	0.707	0.789	0.834	0.925
7	0.582	0.666	0.750	0.798	0.898
8	0.549	0.632	0.716	0.765	0.872
9	0.521	0.602	0.685	0.735	0.847
10	0.497	0.576	0.658	0.708	0.823
11	0.476	0.553	0.634	0.684	0.801
12	0.458	0.532	0.612	0.661	0.780
13	0.441	0.514	0.592	0.641	0.760
14	0.426	0.497	0.574	0.623	0.742
15	0.412	0.482	0.558	0.606	0.725
16	0.400	0.468	0.543	0.590	0.708
17	0.389	0.456	0.529	0.575	0.693
18	0.378	0.444	0.516	0.561	0.679
19	0.369	0.433	0.503	0.549	0.665
20	0.360	0.423	0.492	0.537	0.652
25	0.323	0.381	0.445	0.487	0.597
30	0.296	0.349	0.409	0.449	0.554
35	0.275	0.325	0.381	0.418	0.519
40	0.257	0.304	0.358	0.393	0.490
45	0.243	0.288	0.338	0.372	0.465
50	0.231	0.273	0.322	0.354	0.443
60	0.211	0.250	0.295	0.325	0.408
70	0.195	0.232	0.274	0.302	0.380
80	0.183	0.217	0.257	0.283	0.357
90	0.173	0.205	0.242	0.267	0.338
100	0.164	0.195	0.230	0.254	0.321

Abridged from Table VII of *Statistical Tables for Biological, Agricultural and Medical Research* by R. A. Fisher and F. Yates, by permission of the Longman Group Limited, Harlow.

Appendix C Table of X^2

N	0.90	0.80	0.70	0.50	0.30	0.20	0.10	0.05	0.02	0.01	0.001
							Probability, p				
1	0.016	0.064	0.15	0.46	1.07	1.64	2.71	3.84	5.41	6.64	10.83
2	0.21	0.45	0.71	1.39	2.41	3.22	4.61	5.99	7.82	9.21	13.82
3	0.58	1.01	1.42	2.37	3.67	4.64	6.25	7.82	9.84	11.34	16.27
4	1.06	1.65	2.20	3.36	4.88	5.99	7.78	9.49	11.67	13.28	18.47
5	1.61	2.34	3.00	4.35	6.06	7.29	9.24	11.07	13.39	15.09	20.52
6	2.20	3.07	3.83	5.35	7.23	8.56	10.65	12.59	15.03	16.81	22.46
7	2.83	3.82	4.67	6.35	8.38	9.80	12.02	14.07	16.62	18.48	24.32
8	3.49	4.59	5.53	7.34	9.52	11.03	13.36	15.51	18.17	20.09	26.13
9	4.17	5.38	6.39	8.34	10.66	12.24	14.68	16.92	19.68	21.67	27.88
10	4.87	6.18	7.27	9.34	11.78	13.44	15.99	18.31	21.16	23.21	29.59
11	5.58	6.99	8.15	10.34	12.90	14.63	17.28	19.68	22.62	24.73	31.26
12	6.30	7.81	9.03	11.34	14.01	15.81	18.55	21.03	24.05	26.22	32.91
13	7.04	8.63	9.93	12.34	15.12	16.99	19.81	22.36	25.47	27.69	34.53
14	7.79	9.47	10.82	13.34	16.22	18.15	21.06	23.69	26.87	29.14	36.12
15	8.55	10.31	11.72	14.34	17.32	19.31	22.31	25.00	28.26	30.58	37.70
16	9.31	11.15	12.62	15.34	18.42	20.47	23.54	26.30	29.63	32.00	39.25
17	10.09	12.00	13.53	16.34	19.51	21.62	24.77	27.59	31.00	33.41	40.79
18	10.87	12.86	14.44	17.34	20.60	22.76	25.99	28.87	32.35	34.81	42.31
19	11.65	13.72	15.35	18.34	21.69	23.90	27.20	30.14	33.69	36.19	43.82
20	12.44	14.58	16.27	19.34	22.78	25.04	28.41	31.41	35.02	37.57	45.32
22	14.04	16.31	18.10	21.34	24.94	27.30	30.81	33.92	37.66	40.29	48.27
24	15.66	18.06	19.94	23.34	27.10	29.55	33.20	36.42	40.27	42.98	51.18
26	17.29	19.82	21.79	25.34	29.25	31.80	35.56	38.89	42.86	45.64	54.05
28	18.94	21.59	23.65	27.34	31.39	34.03	37.92	41.34	45.42	48.28	56.89
30	20.60	23.36	25.51	29.34	33.53	36.25	40.26	43.77	47.96	50.89	59.70

When N is greater than 30, use $\sqrt{2\chi^2} - \sqrt{2n-1}$ as a normal deviate.

Abridged from *Statistical Tables for Biological, Agricultural and Medical Research* by R. A. Fisher and F. Yates, by permission of the Longman Group Limited, Harlow.

8 Ecological communities

Soils and woods

If we were to attempt to classify the places where plants and animals live, particularly from the point of view of their relevance for the student of ecology, we would find ourselves thinking at a number of different levels. Coverage could include the whole of the earth's surface where living organisms are to be found from the deepest oceans to the highest mountains (the **biosphere**). Within this huge span, the diversity of habitats is so enormous that in a limited space we could make little of it. Within the biosphere we find a number of large vegetational zones such as the oceans, the tropical rain forests, and the temperate grasslands. These are known as **biomes**. Here again, we are dealing with big geographical areas sometimes widely separated from each other and composed of a great diversity of habitats.

At the next level is the **community**, which we can describe as a group of populations living in the same place. Some of these populations are intimately related to one another through the passage of energy between them along food chains. Such associations of populations are known as **ecosystems** (p. 3). The relationship between other populations is more distant. For instance, that between blue tits eating insects on oak trees and the toadstools living in the soil below amounts to little more than the dropping of bird faeces and the contribution of detritus to the ground when the animals die. The two species can hardly be said to belong to the same ecosystem, but they are both part of the oakwood community.

In this chapter and those that follow we are concerned with the places in which communities live. These are known as **habitats**. For the sake of conciseness we will consider only those occurring commonly in Britain which are reasonably accessible and profitable for the study of ecology.

Soils

As we have seen (p. 98), soil is a highly complex and variable mixture consisting of eroded rock particles, organic debris (humus), air spaces, water and dissolved minerals. Far from being inert, it is 'alive' in that it supports many ecosystems. Soils can be classified for convenience into two kinds.

Throughout the greater part of Britain, the ground tends to be well aerated. Thus any organic matter such as dead leaves which accumulates on the surface, is quickly broken down by aerobic microorganisms to form humus which is incorporated into the thin upper layer (**top soil**). Below the top soil, the proportion of organic material is less and the soil tends to be more compacted. Such conditions are sometimes referred to as **mull** soils. A distinct zonation of top and sub-soils is often not discernible because the calcium compounds that are usually present encourage large populations of earthworms. Their burrowing activities result in mixing of the organic and mineral constituents, the humus present being distributed more deeply. Charles Darwin estimated that an average earthworm community, mainly *Lumbricus terrestris*, numbers about 20 000 ha^{-1}. More recent studies have shown that the distribution of the twenty-five or so species of worms found in Britain is more complex than Darwin thought, up to twelve different species being found on occasions in a single field. Numbers, too, are a good deal greater than was formerly supposed, 200 000 ha⁻¹ being commonplace, while in highly cultivated soils containing an abundance of organic matter and a favourable mineral balance, this figure may be doubled. Such a high density represents a considerable biomass, and the estimates of White [8.1] for several different localities are summarised in Table 8.1.

Table 8.1 Estimates of biomass of earthworms under different types of land use

Land use	Earthworm biomass (kg/ha)
Hardwood and mixed woodland	370–680
Conifers	50–170
Orchards (with grass)	287–640
Pasture	500–1500
Arable	16–760

In Britain, the annual turnover of soil from earthworm casting is prodigious and can attain a level of 14 t ha⁻¹, while in some tropical soils the figure can reach around three times this amount. The effects of all this burrowing are far-reaching and include increased soil drainage, smaller particle size, and a rise in the organic content of the soil due to material being dragged in from the surface.

Fig. 8.1 Section through a heathland soil (New Forest) showing zonation of a podsol. A – peat; B – leached zone; C – pan; D – parent material

On porous soils such as sands and gravels, rainfall causes the minerals, particularly the basic compounds of sodium and potassium, to be washed downwards into the parent substance below. The natural aerobic breakdown of the surface vegetation is inhibited by the formation of acids which cannot be neutralised due to the absence of bases. This results in the build-up of acid peat (semi-decomposed plant material) with a pH of around 4.5. Only a limited number of plant species can tolerate such conditions (see p. 237), which are characteristic of heathlands. Acid soil of this kind is known as **mor**. Over a period of time, fine particles of iron and aluminium compounds are washed out (leached) from the top soil, leaving a pale denuded zone above and, lower down, a compacted impervious layer known as a **pan** where the iron and aluminium compounds accumulate. A section dug through typical heathland (New Forest) is shown in Fig. 8.1 and illustrates the characteristic zonation. Soil that has undergone these changes is said to have been **podsolised** (i.e. converted into a **podsol**).

In low lying areas, water tends to accumulate above the pan and become stagnant, forming a **bog**, in which the anaerobic, acid conditions further inhibit the breakdown of dead plant material. This leads to a shortage of essential minerals, particularly nitrate, and some fascinating adjustments by plant colonists to supplement the meagre supply (p. 241). To farm in such conditions (e.g. on Exmoor) it is necessary to break up the pans by ploughing so as to improve drainage, and to neutralise the acids by the application of lime.

Inhabitants of the soil

The world of the soil is one of darkness, humidity and a reasonably constant temperature, which usually includes an abundance of humus and dissolved minerals. Its colonists can be classified into three distinct groups:

(i) Inhabitants of the soil water
In many ways, the capillary water that occupies the spaces between the soil particles and the water enveloping the particles themselves provide an ideal medium for the existence of microorganisms. Estimates suggest that a single gram of well manured soil may contain as many as 5×10^9 bacteria representing a mass of more than 1.5 tonnes of living matter per hectare. Protozoa also abound, at densities approaching $1 \times 10^6 \, g^{-1}$ of soil and distributed roughly in the proportions of 28×10^4 amoebae, 77×10^4 flagellates, and 1×10^3 ciliates. Many other minute members of the water fauna, such as turbellarians (flatworms), rotifers (wheel animalcules), and tardigrades (water-bears), also occur in huge numbers. Some idea of their density can be gained from that of the nematodes (round-worms) which have been estimated to be present in fertile grassland at a maximum level of about $30 \times 10^6 \, m^{-2}$ with a mass of nearly 18 grams. Comparable figures for the other small organisms are equally staggering [8.1].

(ii) Occupants of the air spaces
From an agricultural standpoint it has been suggested that an 'ideal' soil should consist of approximately 50 per cent solid matter (parent substance and humus), 25 per cent water and dissolved minerals and 25 per cent air [8.2]. Soil is, in fact, a highly variable mixture but air is always a significant part of its composition. The colonists of air spaces are all organisms with a high sensitivity to desiccation.

Many never leave the environment of the soil at all, while others only venture out after nightfall. The community is largely made up of anthropods, the majority being insects, and some typical examples are illustrated in Fig. 8.2. These include isopods (Crustacea), mites (Arachnida), springtails and beetles (Insecta), the springtails being estimated to constitute on occasions as much as 90 per cent of the total numbers. Some idea of the density of these populations can be gained from the estimates of Kevan [8.3], who suggests that one square metre of fertile soil sampled to a depth of 100 mm may include 13×10^4 springtails, 83×10^3 other insects and their larvae, and more than 3×10^5 mites. As for their food, while a few of the larger species are carnivorous, the majority are detritivores. For further information, see the excellent account by Leadley Brown [8.4].

(iii) Burrowers
The principal burrowers in soil belong to three main groups, mammals, annelids, and arthropods. The mammals include tunnellers like the mole and excavators such as the rabbit and badger. Annelids are represented by the various species of earthworms and mention has already been made (p. 149) of their mode of life and important influence on the nature of the soil. The larger arthropods (see Fig. 8.2) such as woodlice, beetles, centipedes, and millipedes are found near the surface and are highly active, emerging onto the surface after dark when conditions in the microclimate are cool and humid and the danger of desiccation is therefore low. It

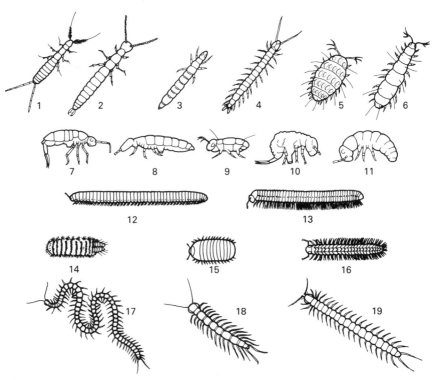

Fig. 8.2 Some typical soil arthropods, 1–2 diplurans (order Diplura); 3 proturan (order Protura); 4 symphylid (class Symphyla); 5, 6, 9 pauropods (class Pauropoda); 7, 8, 10, 11 springtails (order Collembola); 12–16 millipedes (class Diplopoda); 17–19 centipedes (class Chilopoda) (after Kevan)

is these species that are readily captured in pitfall traps (p. 78). As might be expected from their mobility, many of them are carnivores, e.g. centipedes and some beetles.

The litter layer

Covering the surface of the soil and particularly evident in areas of high leaf-fall such as woodland, is a zone of semi-decomposed plant material known as the **litter layer**. Under trees whose leaves decompose slowly, this can be 50 mm deep or more. The litter has an important influence on the soil below, as it provides a continuing source of humus and helps to maintain the level of humidity at the surface. Although subject to greater environmental fluctuations than the soil underneath, particularly in temperature and humidity, it is nonetheless a favoured habitat for many small arthropods such as springtails.

Moisture is no doubt the main abiotic factor influencing animal distribution in the litter layer and typical cryptozoic species such as woodlice, centipedes and millipedes show a negative taxis to dry air and light. This can easily be tested in a choice chamber. In woodlice, high humidity (possibly associated with low temperature) tends to reduce activity and hence prevents the animals from wandering too far from their optimum conditions. This may also account for the large local aggregations of woodlice and other arthropods that frequently occur underneath rotting logs and large stones. It has been shown that different species vary in their degree of drought tolerance. Like their counterparts in the top soil, most of them only venture out after dark but even then, they can only withstand the reduced humidity for relatively short periods. Among woodlice, the most tolerant species belong to the genus *Armadillidium*, so called on account of their capacity to roll up into a ball like a miniature armadillo. Curiously, these litter layer animals tend to become positively phototactic in dry air and this leads to a kind of do or die behaviour pattern. In the event of their habitat becoming too dry to sustain them, they tend to react by wandering about in the open until they either find a more suitable site or perish in the attempt.

Some topics for investigation

1 In what respects does the zonation of two (or more) different soils, e.g. cultivated and uncultivated, vary? Can you account for any differences observed?

2 In a similar study to 1 above, how do edaphic factors such as water content, humus, and pH vary? Can you relate these to colonising plant populations, e.g. are they calcicole or calcifuge? How do the different zones vary from one another?

3 Using the formalin method (p. 81), sample earthworm populations inhabiting two different kinds of ground, e.g. cultivated and uncultivated. Identification of species can be difficult and time-consuming, so the number collected may be a misleading parameter. Would biomass be better? If required, different species can be identified using a specialist key [8.5].

(a) How does earthworm density (mass m^{-2} treated) vary between localities? Can you account for any differences?

(b) To what extent do successive samples from the same area vary? Is there any discernible distribution pattern? Are the differences between samples statistically significant?

(c) Assuming an average burrowing depth (say 50 cm), calculate the volume of soil with a surface area of 1 m². Estimate the humus present assuming uniform distribution (is this justified?). Compare this with the estimated biomass of earthworms. Is it likely that humus availability could be a limiting factor controlling the size of earthworm populations? What other factors might be concerned?

4 What variations occur in the density of small arthropod populations, e.g. springtails, in the litter layer, top soil and sub-soil? Can you suggest reasons for any differences?

Do the flotation and Tullgren funnel methods of extraction (p. 82) give comparable results? If not, what are some of the reasons for any differences?

5 Examine and compare the water fauna from different soils and levels using the Baerman funnel method of extraction. Can you suggest any reasons for the differences observed, both in numbers and diversity of species?

6 Using pitfall traps (p. 78) compare the activities of the larger arthropods inhabiting top soil and litter in two different localities, e.g. a wood and a garden.

(a) Are the species and their numbers similar in the two places? If not, can you suggest any reasons for the differences?

(b) Mark the animals captured with dots of cellulose paint (p. 90), then release them where they were caught. Do the different species have distinct distributional patterns? To what extent can the populations be said to be separate or to overlap? What is the range of movement of the different species and how does it vary with climatic conditions?

Woods

It is doubtful if any of the woodland existing in Britain today can be called 'natural' in the sense that it has been unaffected by the activities of man. However, trees by their very nature, particularly the limited dispersal of their fruits and seeds, seldom exist alone but usually occur in aggregations. In spite of human interference, woodland now exhibits most of the ecological characteristics that it has possessed throughout its history and because of its variety, it is the most complex habitat existing on land.

Classification of woodland

The growth of trees to form woods represents the climax stage of an ecological succession (p. 67). Which species will predominate and to what extent, is determined by a number of environmental factors of which soil is the most important. The other factors include exposure to wind, the range of temperature variations, and whether conditions are favourable for germination and the early stages of growth, particularly whether there is sufficient light and an absence of herbivore predators.

One way of classifying woods is by the predominant plant species composing them; thus we refer to oakwoods, beechwoods, and alder thickets. This procedure is imprecise, for it frequently happens that woods are of mixed species and the

Fig. 8.3 Scots pine, *Pinus sylvestris*, on typical heathland among ling, *Calluna vulgaris*

only feature the trees have in common is that they are all deciduous. In general, oak prefers soils with a high content of clay, while chalk and limestone are favourable to beech, which thrives in alkaline conditions, although it can also tolerate lime-free situations provided it is not competing with oak. Ash is also a tree with strong calcicole tendencies and so favours soils with a high pH (7.5 or more). In damp conditions, such as water meadows, willow is frequently prominent, while, by the sides of streams, alder often takes over. Again, the acid conditions of heathland (pH around 4.5) favour birch and conifers such as the Scots pine, *Pinus sylvestris* (Fig. 8.3).

Stratification of woodland

A more precise method of classifying woods is in terms of their internal composition and the vertical zonation of the different layers of vegetation. The extent to which such zonation (**stratification**) occurs varies considerably from one ecological situation to another, but in the most diverse communities (e.g. mixed deciduous woodland) four layers are usually discernible. These are:

(i) **the tree layer** which forms a canopy over the rest of the vegetation below;
(ii) **the shrub layer**, a variable quantity ranging from brambles about 1 m high to hazel extending to 10 m or more;
(iii) **the field layer** consisting of a wide range of herbaceous annuals and perennials which vary greatly in their diversity and numbers from one situation to another;
(iv) **the ground layer** dominated by small plants such as mosses, liverworts and lichens. Here, too, belong the numerous fungi (many of them basidiomycetes) whose fruiting bodies (toadstools) are a feature of woodland in autumn.

Fig. 8.4 Typical wood of beech, *Fagus sylvatica*, in spring. During summer the dense canopy excludes most of the light inhibiting the growth of shrub and field layers

Tree layer

The fact that the tree layer is so easily discernible in most British woodlands is a commentary on the paucity of species occurring here. In the tropical rain forests of Central and South America (some of the densest on earth) many trees of different species, heights, and ages all grow together, forming a canopy of great complexity. The most important ecological effect of the tree canopy is in controlling the amount of light that can penetrate below. Thus, in beechwoods, the dense foliage shuts out much of the illumination during the important growth period from April to October with the result that shrub and field layers are frequently absent (Fig. 8.4). The same is true of coniferous forests (Fig. 14.5 p. 276) such as pine or spruce (frequently planted by man) but here the situation is more extreme since the trees are evergreens and therefore retain their leaves all the year round. By contrast, mixed deciduous woods are more open (particularly if they have been artificially thinned to promote strong growth by reducing competition) and the high illumination encourages dense growths of shrubs and herbs (Fig. 8.5).

Shrub layer

This ranges from small woody plants such as bramble, *Rubus fruticosus*, and privet, *Ligustrum vulgare*, to spreading tree-like plants such as hazel, *Corylus avellana*, and holly, *Ilex aquifolium*, which can reach a height of 7 m or more. Like the trees above them, the shrubs can also exert a strong shading effect, as is well illustrated by a mature hazel coppice in mid-summer which can give the impression of almost total darkness inside.

Field layer

For the herbaceous annuals and perennials living below the shrub layer, it is often a race against time to ensure flowering and the setting of seed before light shortage

Fig. 8.5 Mixed deciduous woodland. The abundance of light promotes a dense growth of shrubs and herbs

intervenes. This is particularly true of species which rely on insect pollination such as the wood anemone, *Anemone nemorosa*, illustrated in Fig. 8.6. An outcome of this situation is that a distinct flowering sequence is usually observable among plants of the field layer from the early spring onwards. First to appear in February are the primrose, *Primula vulgaris*, and celandine, *Ranunculus ficaria*, then the wood anemone, *Anemone nemorosa*, in March. Bluebells, *Endymion nonscriptus*,

Fig. 8.6 Wood anemone, *Anemone nemorosa*. An early flowering species of the field layer in woodland

flower in April and continue into the deep shade period of early May, while plants such as meadowsweet, *Filipendula ulmaria*, and valerian, *Valeriana officinalis*, are able to flower successfully in the dim light of summer from June onwards. It should be noted that the flowering sequence characteristic of the field layer plants in spring is not a succession in the generally accepted sense (see p. 67). Rather it represents a subtle sequential adjustment of the behaviour pattern of different species in response to changing environmental conditions. Living apparently successfully within the field layer are sometimes to be found plants usually associated with areas of high light intensity. An example is the rosebay willowherb, *Epilobium angustifolium* (Fig. 1.3), a colonist of forest clearings, particularly where burning has occurred. However, more careful observation will show that although these plants succeed in growing and indeed extend their range by underground rhizomes, they are incapable of flowering and must therefore have arrived as seeds from elsewhere. The implications of this will be discussed further in the next section.

Ground layer
This consists of a wide variety of lowly plants such as mosses, liverworts, and lichens which abound in areas where there is sufficient moisture, for example the base of tree trunks, the banks of ditches, on rotting timber and so forth. Nearly all these species are able to tolerate a low level of light intensity. Another important component of the ground layer is invisible for most of the year – the fungi. These are mostly basidiomycetes, whose spore-bearing bodies (toadstools) are a feature of woodland in October. Many of them are saprophytic, some form symbiotic associations with trees and shrubs (mycorrhiza), while others such as the honey fungus, *Armillaria mellea* (Fig. 8.7), are serious parasites of many woodland trees such as oak.

Fig. 8.7 Honey fungus, *Armillaria mellea*, growing on oak, *Quercus robur*

Adaptation to light intensity

One of the themes running through the previous account is that plant species vary greatly in their tolerance of illumination and their capacity to utilise such light as is available. In order to survive, a plant must balance its energy budget so that production equals or exceeds consumption, i.e. the rate of respiration is balanced by that of photosynthesis. There are two problems here which have been well illustrated by Ashby [5.6] and are explained in Fig. 8.8.

(i) In order to achieve a balance between the rates of photosynthesis and respiration (**compensation point**) plants need to make the best use of the available light.
(ii) The time taken to attain the compensation point (**compensation period**) must be short enough to match the duration of illumination in a particular habitat.

The compensation periods of plants can be measured in the laboratory [8.6], from which it appears that the performance of different species varies widely. Fig. 8.8 shows that shade plants such as bluebells and anemones are characterised by efficient photosynthesis at low light intensities, coupled with a relatively low level of respiration. Sun plants with a high compensation point are typified by species with vigorous growth, such as weeds and colonists of downland, which usually indulge in continuous growth throughout the summer until such time as flowering occurs. The fact mentioned earlier that rosebay willowherb, *Epilobium angusti-folium*, a typical sun plant, can colonise woodland but is unable to flower, can be explained by assuming that although the compensation point may be reached, additional food reserves needed for flowering cannot be produced. Many shade plants in woodland have underground storage organs, for example the bulbs of bluebells, and these tend to obscure the pattern of growth in promoting an outburst in spring after which few further leaves are added before the foliage begins to die down. The distinction between sun and shade plants is less precise than appears at first sight, for within the whole range of species there is infinite variation. Moreover, the concept of the compensation point as a level of light intensity at which an energy balance is achieved fails to meet such requirements as the need to form food reserves for overwintering and nightfall. There is also the

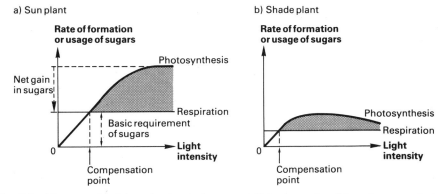

Fig. 8.8 Photosynthesis/respiration balance and light intensity. Note that a typical shade plant has not only a lower compensation point than a sun plant, but it cannot make such good use of higher light intensities (after Ashby)

necessity for the plant to adjust to varying daylength during the period of photosynthetic activity. One way of bringing together all these variables is to express the compensation point as the percentage of full daylight needed by a species *in the field* in order to survive (including reproduction). The term **extinction point** is sometimes used for this purpose and is defined as the lowest percentage of full daylight in which a species occurs under natural conditions. It is an easy parameter to apply in practice, for all that is needed is a series of light readings in places where a species exists. Signs that a species is nearing the extinction point will be inadequate growth or a failure to flower, as in the rosebay willowherb in woodland mentioned earlier. Incidentally, reliable measurements of light intensity are not as easy to make as it might appear, and to obtain these certain precautions are necessary (see p. 106).

Animal colonists of woodland

With such a diverse range of plants comprising woodlands, it is hardly surprising to find that the range of ecological niches available for animals is equally variable.

The most obvious product of plants during summer is leaves, and these provide food for a host of small insects and their larvae – leaf miners, sap feeders, gall formers, and defoliators of several different kinds. The relationships of some of these niches are summarised in Fig. 8.9.

Other parts of plants such as wood, flowers, fruits, seeds, bark, and roots are also important sources of food for many animal populations. These include wood borers, nectar feeders, seed eaters, under-bark communities, and herbivorous mammals. Browsers such as rabbits play an important part not only in devouring the established herbs of the field layer but also in destroying young seedlings. Their influence in inhibiting the regeneration of trees from seed is only exceeded by the effects of human trampling, which, in much visited woodlands, virtually inhibits the growth of young woody plants except in specially protected areas set aside for the purpose. Some typical niches and their relationships in woodland based on parts of plants other than the leaves are summarised in Fig. 8.10.

The soil of woodland is usually rich in humus from fallen leaves and supports large populations of detritivores such as earthworms and springtails. The latter also often abound in the litter layer, which can be 50 mm or more thick in beech and oakwoods. The field and shrub layers support great numbers of insects, which can be detached by beating into a tray (p. 77), while the rough bark of trees such as oak supports a peculiar insect fauna of its own. Dominating the food webs of woodland is a wide assortment of bird species – herbivores, insectivores, and larger carnivores. These can provide a study on their own and suggestions were made earlier (p. 91) regarding methods for estimating their numbers. For those wishing to pursue the subject further, reference should be made to the account by Yapp where his methods are developed in the context of different kinds of woods [8.7].

Beginning an ecological study in woodland can be a rather daunting experience, not only on account of its complexity but also because its composition can change radically within a few yards. Thus, an oak wood can give way to a predominance of beech or ash within a short distance, indicating that a layer of clay has been superseded by a seam of chalk or limestone. This only serves to underline the point made several times already that when beginning an ecological investigation it is advisable to choose a small area to study, at least as a starting point.

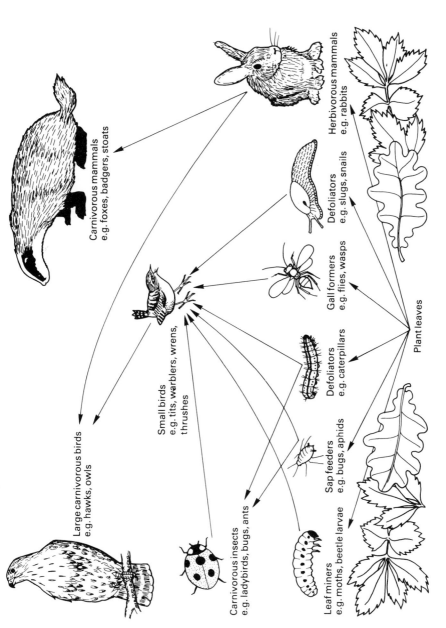

Fig. 8.9 Some ecological niches and food relationships in woodland, based on plant leaves (after Neal)

Carnivorous mammals
e.g. foxes, badgers, stoats

Herbivorous mammals
e.g. rabbits

Large carnivorous birds
e.g. hawks, owls

Small birds
e.g. tits, warblers, wrens,
thrushes

Carnivorous insects
e.g. ladybirds, bugs, ants

Defoliators
e.g. slugs, snails

Gall formers
e.g. flies, wasps

Defoliators
e.g. caterpillars

Sap feeders
e.g. bugs, aphids

Leaf miners
e.g. moths, beetle larvae

Plant leaves

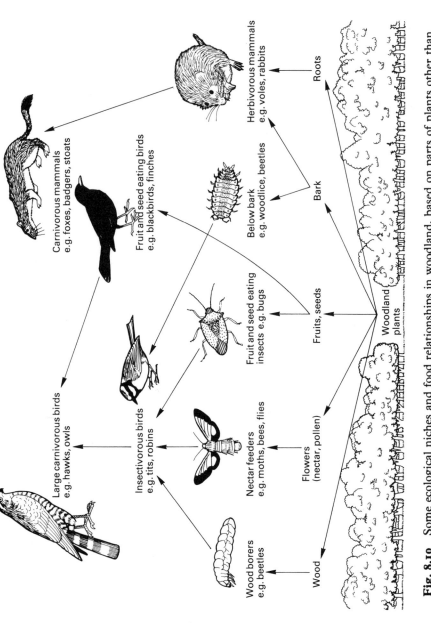

Fig. 8.10 Some ecological niches and food relationships in woodland, based on parts of plants other than their leaves (after Neal)

Some topics for investigation

1 How does the stratification of woodland vary with different tree canopies, e.g. oak and beech?

2 What is the distribution pattern of plant species in the field layer in different situations? To what extent does any particular physical factor, e.g. light, influence distribution?

3 In studying plant distribution, what is the pattern of coverage (percentage cover) achieved by different species? Is it possible to work out an index of diversity (p. 22)? What, if anything, do these parameters tell us about the age of the wood and any external influences to which it may have been subjected?

4 How does the composition of the shrub, field, and ground layers in a piece of undisturbed woodland compare with that of an area that has been recently cleared?

5 In an area that has been recently cleared, what evidence is there of a pattern of succession? As a long-term investigation using permanent quadrats (p. 86): how does succession in a cleared and uncleared area compare?

6 In a comparison of the distribution of sun and shade plants in a piece of woodland, is there any evidence of sun plants growing in the shade and nearing their extinction point?

7 What is the range of organisms, both plant and animal, occupying different parts of a tree such as an oak? In what ways could they be related ecologically? For a stimulating introduction to such a study see Darlington (1972) [8.8].

8 To what extent can the insects obtained by beating from the tree, shrub, or herb layers be related to a particular ecosystem? For instance, evidence may be available regarding their diet (judged by behaviour and mouthparts) and their relative numbers.

9 Many plants react to the invasion of their tissues by insects and other arthropods by producing galls. A typical example is the spangle galls which appear on oak leaves in late summer, which are caused by the larvae of the gall wasp, *Neuroterus*, of which four species are involved (Fig. 8.11). In a sample

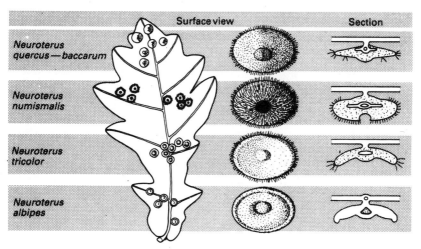

Fig. 8.11 Spangle galls on oak leaves containing larvae of four different species of the wasp, *Neuroterus* (from Darlington)

of oak leaves, what is the pattern of distribution of each species? To what extent do their distributions (a) overlap, (b) coincide. How do the results of sampling vary for different trees and different woods? For a useful introduction to this problem see the account by Lewis and Taylor [8.9] and the excellent key to the identification of galls by Darlington [8.10].

10 To what extent does natural regeneration from seed occur in a wood dominated by a particular tree species? Which factors appear to be promoting seedling growth and survival, and which are having an inhibiting effect?

11 Woodland provides excellent opportunities for studying the flow of energy in an ecosystem, as Wells [8.11] has shown. The course he has devised covers (a) the intake and storage of energy by plants and small animals, (b) energy utilisation in growth and respiration, (c) energy consumption by herbivores. Full details of the methods used are given in the article quoted above. These could easily be adapted for other ecosystems and would provide good material for project work.

Summary

1 The ecology of organisms can be studied at a number of different levels ranging from the whole of the earth's surface (biosphere), to large vegetational zones (biomes), to small groupings of populations (communities), to populations closely associated by virtue of food chains (ecosystems). The places where communities live are known as habitats.

2 Mull soils are characteristic of the greater part of Britain. They are well aerated and organic matter accumulating near the surface is quickly broken down by microorganisms.

3 Within such soils the density of earthworms varies greatly. Well aerated loams with an abundance of organic matter and a favourable mineral balance, particularly of calcium (Ca^{2+}) ions, support the largest populations.

4 Sands and gravels are subject to erosion of bases and the build-up of acid conditions with much semi-decomposed organic matter (peat). Such leached soils are known as mor and are said to be podsolised.

5 When mor soils are low lying, water tends to accumulate, forming bogs.

6 The animal colonists of soil can be grouped as inhabitants of the soil water, occupants of air spaces, and burrowers.

7 The litter layer, which consists of semi-decomposed plant material, is a favourite habitat of many small soil arthropods.

8 It is doubtful if any woodland existing in Britain today has been unaffected by human activity and interference.

9 Woods can be classified by their dominant tree species, whose distribution is sometimes closely related to soil conditions.

10 A more precise method of classifying woodland is by its stratification into tree, shrub, field, and ground layers.

11 The spatial distribution of woodland plant species is closely related to light intensity. The compensation point provides a measure of their adaptability, the limits being determined by the extinction point.

12 Woodland supports a great diversity of animals, often at a high density. No portion of a plant is immune to attack by herbivores. These in turn provide food for a variety of carnivores, many of them birds and insects.

13 The soil of woodland is usually rich in humus and supports large populations of detritivores, particularly small arthropods.

14 Sudden variations in the nature of the soil can result in correspondingly rapid changes in the composition of woodland.

References

8.1 White, R. W. (1979) *Introduction to the Principles and Practice of Soil Science*, Blackwell

8.2 Russell, E. J. (1957) *The World of the Soil*, Collins

8.3 Kevan, D. K. McE. (1961) *Soil Animals*, Witherby

8.4 Leadley Brown, A. (1978) *Ecology of Soil Organisms*, Heinemann

8.5 Edwards, C. A. and Lofty, J. R. (1967) *Biology of Earthworms*, Chapman and Hall

8.6 Baron, W. M. M. (2nd edn 1967) *Organization in Plants*, Arnold

8.7 Yapp, W. B. (1962) *Birds and Woods*, Oxford

8.8 Darlington, A. (1972) *The World of a Tree*, Faber

8.9 Lewis, T. and Taylor, L. R. (1967) *Introduction to Experimental Ecology*, Academic Press

8.10 Darlington, A. (1968) *Plant Galls*, Blandford

8.11 Wells, P. (1982) 'An ecological energetics field course for sixth forms', *Journal of Biological Education*, **16**(4), 265–74

Ecological communities

Downs and dunes

As far as we can tell, virtually all the areas of grassland existing in Britain now were at one time forest. Some of these were cleared a long time ago (perhaps about 2500 B.C.), others are of more recent origin. Today, most of our grassland is used for agricultural purposes of one sort or another and is periodically ploughed, reseeded, treated with fertilisers, and sprayed with selective herbicides. Its quota of plants and animals is therefore largely controlled by man [9.1].

Downs

The only remaining areas that can lay any claim to being 'natural' in that they are at least partly independent of human interference are our downlands, which are located mainly in southern England, Yorkshire, and Derbyshire and are associated with chalk and limestone. Chalk is an almost pure form of calcium carbonate, while limestone is more impure. Both produce soils which are alkaline, pH ranging from about 7.4–8.5 and are colonised by a diverse range of lime-tolerant plant species (calcicoles).

Another feature of downland, particularly chalk, is that the surface soil tends to be shallow, so that water and minerals are concentrated in the first few centimetres. A typical analysis made by Salisbury [9.2] is shown in Table 9.1.

Table 9.1 Analysis of a chalk downland soil (after Salisbury)

pH	Depth (mm)	Water content (% dry soil)	Calcium carbonate (%)	Organic material (%)	Ratio of water : organic
7.4	25	103.0	3.0	35.05	2.9
7.5	50	66.4	6.7	23.0	2.9
7.6	75	55.6	7.8	19.04	2.9
7.6	100	54.4	9.9	18.9	2.8
7.6	125	41.3	23.0	—	

The tendency to concentrate moisture at the surface means that downland soils (particularly on chalk) dry out rapidly during warm weather in the summer. This effect is greatly accentuated where the slope of the ground is steep. The influence of these conditions on the plant populations is threefold:

(i) shallow rooted plants such as grasses achieve most of their vegetative growth during the early part of the year while conditions are most favourable. Typical downland colonists are sheep's fescue, *Festuca ovina*, various species of bent, *Agrostis*, and the tor grass, *Brachypodium pinnatum*;

(ii) many perennials form two types of rooting system, a shallow network and roots extending much deeper into the sub-soil. A typical example is the milkwort, *Polygala calcarea*, one of the most attractive and ubiquitous species associated with downland (Fig. 9.2);

(iii) on the sloping ground often characteristic of the edge of downland, the problem of drought is accentuated, and a number of the successful plant colonists exhibit xeromorphic characteristics (structural adaptations for retaining water) (see also p. 172). Typical among such species is the wall pepper, *Sedum acre*, which also thrives on old limestone walls (Fig. 13.4 p. 255).

Succession on downland

As we have seen, the climax vegetation from which all downland has been derived is woodland, frequently oak. Once cleared, regeneration has been kept in check by the natural browsing pressures of herbivorous mammals, particularly rabbits, and more recently, by the pasturing of sheep. The situation is thus a dynamic one: while grazing pressure is high the preservation of downland is achieved, but should it be relaxed, rapid and profound successional changes can be expected.

The characteristic features of both chalk and limestone downland are a closely cropped springy turf supporting the most diverse range of plants found in any habitat in Britain. Many of these belong to the pea family (Papilionaceae), for example, the bird'sfoot trefoil, *Lotus corniculatus* (Fig. 9.1), lady's fingers, *Anthyllis vulneraria*, and horseshoe vetch, *Hippocrepis comosa*. Many species, for example the rock rose, *Helianthemum nummularium*, and the self-heal, *Prunella vulgaris* (the latter a pernicious weed in lawns), show a remarkable resistance to the effects of browsing and trampling. One of the reasons why this latter plant, and other species like it, spread so readily is that small detached fragments of their leaves have the ability to form roots when in contact with most soil, giving rise to new plants. The fact that the range of a colony of the rare cut-leaved self-heal, *Prunella laciniata*, was found to have increased considerably after being drilled on

Fig. 9.1 Bird'sfoot trefoil, *Lotus corniculatus*. A typical colonist of downland and an important food for insect and mollusc herbivores

the previous year for a fortnight by a platoon of soldiers, may well be explained in this way. Incidentally, there is a moral here for gardeners who choose to mow weedy lawns without a grass box!

Another feature of the downland flora is the development of a rosette habit of growth in which the leaves remain close to the ground. Reference has already been made to the milkwort, *Polygala calcarea* and Salisbury has estimated that rosette plants, including a great diversity of different families, comprise some 35 per cent of the entire chalk flora. Many of the orchids that are a feature of downland have a rosette structure in their non-flowering condition, a typical example being the spotted orchid, *Dactylorhiza fuchsii* (Fig. 9.3). This pattern of growth could well represent a response to the selective pressure of grazing which has evolved over a considerable period of time.

Like all animal populations, those of herbivorous mammals are subject to periodic fluctuations in numbers. Usually, these variations are relatively small, so that a lack of cropping one year is more or less counterbalanced by an excess the next. However, on occasions more drastic checks may occur, such as the myxomatosis epidemic of the 1950s which is estimated to have exterminated more than 90 per cent of the rabbits in Britain. The sudden reduction of grazing pressure had remarkable consequences and populations of strongly growing grasses such as the tor grass, *Brachypodium pinnatum*, which had previously been kept in a lawn-like condition, rapidly assumed the appearance of a hay field. In some places, at least 75 per cent of the low-growing downland species of flowering plants had been eliminated within two years.

The natural successors to such conditions are shrubs such as hawthorn, *Crataegus monogyna*, and bramble, *Rubus fruiticosus*. Unless removed artificially, hawthorn rapidly forms thickets (Fig. 14.11 p. 289). These are attractive to nesting birds such as greenfinches, which eat the berries and pass the seeds out in their faeces, thereby acting as an effective dispersal agent. Several species of tree such as the ash, *Fraxinus excelsior*, and the turkey oak, *Quercus cerris*, can compete successfully with hawthorn and these often mark the next stage in succession, particularly on limestone downland.

Animals on downland

The abundance of calcium compounds in downland soils makes them particularly attractive to chalk-loving animals such as gastropod molluscs. In many areas the small black and white shells of *Helicella itala* abound on short turf, and these provide the food for the larvae of the beetle, *Lampyris noctiluca*, which are commonly known as glow-worms. Thrush anvil stones are a frequent occurrence (Fig. 9.4) and bear evidence of the diet of the song-thrust, *Turdus philomelos*. The diet includes numbers of the larger snails such as the white-lipped snail, *Cepaea hortensis* (Fig. 9.5). This snail often forms dense populations, feeding for preference on dead vegetation and becoming active if there is a dew or after a shower of rain.

Among the downland animals observable in daytime, the most conspicuous are the butterflies, and these pose some interesting ecological problems. The commonest species belong to two families, the Lycaenidae (blues) and Satyridae (browns). The latter are particularly convenient for study as they fly in any weather provided it is reasonably warm. Among the blues are the common blue, *Polyommatus icarus*, the small blue, *Cupido minimus*, the adonis blue, *Lysandra*

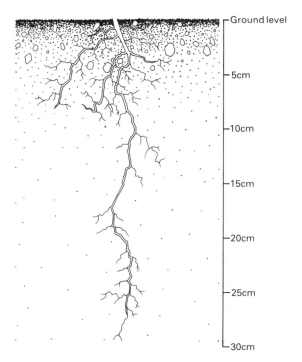

Fig. 9.2 Rooting system of milkwort, *Polygala calcarea*, on downland. Note the deep and shallow roots (after Salisbury)

Fig. 9.3 Spotted orchid, *Dactylorhiza fuchsii*. A common species with a rosette habit of early growth

Fig. 9.4 Thrush anvil stone on downland with shells of the garden snail, *Helix aspersa*, brown-lipped snail, *Cepaea nemoralis*, and white-lipped snail, *C. hortensis*

bellargus, and the chalkhill blue, *L. coridon*; the latter, as its name implies, is confined to the chalk. The larvae of all these species feed on a variety of low growing leguminous plants. The satyrines are all grass feeders when larvae and include the hedge brown, *Pyronia tithonus*, which is confined to hedges at the periphery of downland; the ringlet, *Aphantopus hyperanthus*, which occurs at the edge of downland and which overlaps the hedge and meadow browns to some

Fig. 9.5 White-lipped snail, *Cepaea hortensis*, on downland. Note the variation in the degree of banding on the shell

extent; the meadow brown, *Maniola jurtina*, and marbled white, *Melanargia galathea*, colonists of open downland; and the speckled wood, *Pararge aegeria*, which occurs at the edges of woods and in open spaces inside them. The downland insect community also includes a number of day-flying moths, typical among them being the six-spot burnet, *Zygaena filipendulae*, whose cocoons are often prominent on grass stems in early summer.

The ecosystems occupying grass tufts include many of the larvae of the species already mentioned, other adult herbivores such as slugs and snails, and a variety of carnivores including ants and some beetles. During the day these populations remain deep down among the grass stems, no doubt an adaptation enabling them to avoid desiccation and the notice of predators such as birds. At night, they become active and many individuals can be collected successfully with a sweep net as they ascend to the surface of the vegetation.

A peculiar feature of a few species of downland plants is their ability to emit the poisonous substance hydrogen cyanide when cut or bruised. These **cyanogenic** forms occur in clover, *Trifolium repens* and bird'sfoot trefoil, *Lotus corniculatus*, and no doubt represent a protective device against browsers such as snails. The presence or absence of cyonogenesis is easily detectable in the field using the sodium picrate paper test and this is an aspect of downland ecology that would repay further study [9.3].

Some topics for investigation

1 How does downland plant distribution in an area in full sun differ from that in the shade, e.g. below hawthorn? Parameters could include (a) the number of species and their identity, (b) the percentage coverage of each species, (c) the diversity of species.

2 What are the effects of trampling on downland plants? How does plant distribution vary (a) in the middle of a pathway, (b) at the side, (c) in the adjoining grassland? Is it possible to construct an index of resistance to trampling for different plant species?

3 What information on downland ecosystems is provided by (a) beating the insects from bushes such as hawthorn, (b) sweeping grass tufts by day, (c) sweeping grass tufts by night (preferably when conditions are fairly warm and humid)?

4 From observations of thrust anvil stones (p. 168), how does the occurrence of the shells of snails such as *Cepaea nemoralis* and *C. hortensis* provide evidence of differential selection by birds? Features for comparison of predated and living individuals could include (a) size, (b) banding on shells, (c) background colour of shells (usually yellow in *C. hortensis* but brown, yellow or pink in *C. nemoralis*). How do the results from downland compare with those in neighbouring woodland? Does the selective advantage/disadvantage of a particular characteristic vary seasonally?

5 What is the pattern of distribution of the cyanogenic forms of clover or bird'sfoot trefoil (the sodium picrate test for cyanide can easily be conducted in the field [9.4]). Does this appear to be related to any physical factors such as the slope of the ground (dryness) or altitude (frost pockets)? Is there any evidence of selective predation on plants by slugs and snails (e.g. nibbling leaves of some plants and not others)?

6 What is the sequence of succession on a particular area of downland (most stages are usually present if only in small numbers)? What is the successional climax? Is there any evidence of (a) the rate at which succession is taking place when uninterrupted, (b) the results of management in controlling succession? What effects is plant succession likely to have on the animal populations of downland?

7 Satyrine butterflies such as the meadow brown and marbled white are excellent material for the mark-release-recapture method for estimating density (see p. 94). Choose a downland population, bounded if possible by a physical barrier such as a wood on one or more sides. What is the density of the population expressed as numbers per square metre (preferably at different times from late June onwards)? (Males and females should be regarded as separate populations.) Is the sex-ratio constant? To what extent do marked individuals fly out of the area of study and individuals marked outside it move into it?

Dunes

Where the land meets the sea an interaction frequently takes place which results in a highly specialised type of habitat, the basis of which is sand blown shorewards by the wind. The sand piles up into unstable barriers which may reach a height of 40 metres or more. Although sand dunes tend to be formed in areas where the wind is onshore, secondary influences such as variations in wind pattern at different times of the year and erosion by the sea may play an important part in determining their eventual size and outline. Thus, while dunes may assume a somewhat variable appearance, there are certain characteristics that are fundamental to them all and it is with these that we shall be concerned in this section [9.5].

Plant succession on dunes

Dunes are ideal places to study the process of plant succession because:

(i) the habitat is clearly defined and easily recognisable;
(ii) the sequence of events, although somewhat variable in their pattern, is nonetheless fairly easy to comprehend;
(iii) the number of plant and animal species involved is relatively small.

In general, the successional sequence (**psammosere**) passes through several phases which follow one another but are not always clear-cut.

Phase 1   Young dunes
The earliest stages of colonisation consist of sand and other dry material being blown shorewards by the wind and being arrested by plants already growing there, usually well clear of the high tide mark. Such species are generally adapted for dry conditions (**xerophytes**) and exhibit xeromorphic characteristics such as a thick cuticle on their leaves and the development of fleshy water storage tissue. They are also able to withstand variable concentrations of salt (**halophytes**, see p. 228). Typical species are the sea holly, *Eryngium maritimum* and the sea spurge, *Euphorbia paralias* (Fig. 9.6).

(a)

(b)

Fig. 9.6 (a) Sea spurge, *Euphorbia paralias*, (b) sea holly, *Eryngium maritimum*; two pioneer species in the formation of sand dunes

Phase 2   Partially fixed dunes (yellow dunes)
The banking up of sand results in the formation of mounds, and these are mainly colonised by a number of xerophytic perennials, of which the chief are marram grass, *Ammophila arenaria* (Fig. 9.7) and the sand sedge, *Carex arenaria*, both of which have the capacity for rapid vegetative growth by underground rhizomes. Mosses frequently appear at an early stage in the succession and in common with the roots of the larger plants, have a stabilising effect on the shifting sand. The

Fig. 9.7   (a) A partially fixed (yellow) dune colonised mainly by marram grass, (b) a fixed (grey) dune. The marram grass has been largely replaced by a wide variety of other species, particularly grasses. The bush in the foreground is sea buckthorn, *Hippophae rhamnoides*

commonest of these mosses is *Tortula ruraliformis*, a species whose distribution is confined to sand dunes. At this stage colonisation by plants is incomplete and the dune is said to be *partially fixed*, large patches of sand still being in evidence. These give the dune a yellow appearance, hence the term 'yellow dune'.

Phase 3   Fixed dunes (grey dunes)
The stabilising of yellow dunes is associated with the arrival of more mosses and a wide range of perennial and annual plant species. Of the latter, many survive while

conditions are moist during the winter, early summer, and autumn but once a dry period sets in their shallow rooting systems become desiccated and most of the plants die. Such annual plants are called **ephemerals**. Moreover, marram grass, which flourishes so abundantly on partially-fixed dunes when competition from other species is low, becomes progressively less successful as the process of succession proceeds and competition becomes more intense. The eventual community of plants on fixed dunes can contain a large number of species and Hepburn [9.6] has recorded as many as 250, excluding mosses and liverworts. By this stage in succession most of the pioneer colonists will have been submerged in the sand, and although isolated patches of fast-growing species such as marram grass survive for a time, the majority die, their decomposition contributing a much needed supply of humus.

Habitats behind the dunes

The formation of sand dunes is a dynamic process which moves steadily seawards. One range of dunes may be followed by another, the one nearest the sea affording a measure of protection for the next one. Behind them they leave a fascinating range of habitats each with its characteristic plant and animal populations.

Immediately behind fixed dunes the ground usually falls away and in districts of high rainfall, water accumulates forming moist **slacks**, the dominant plant being the creeping willow, *Salix repens*. Where the ground is low and there is a lack of calcium ions, pools form and the water is usually acid. This is a **dune marsh**, and its ecology closely resembles that of the bogs described in Chapter 12. Under dry conditions, erosion of the bases in the soil by rain and rapid drainage results in the formation of **dune heath**, which resembles closely the heathland occurring inland (p. 236). When the ground is high in calcium ions, a typical downland community can develop and, provided growth of the vegetation is kept in check by grazing (usually by rabbits or sheep), **dune grassland** results. If subjected to the strong sea winds, the habitat tends to be dry in summer and hence favours an unusually high proportion of plants with xeromorphic adaptations, for example, the several species of stonecrop (*Sedum*) with fleshy water storage tissue in their leaves. In the absence of grazing pressure, the process of succession results in bushy scrub, frequently dominated by sea blackthorn, *Hippophae rhamnoides*, elder, *Sambucus nigra*, and privet, *Ligustrum vulgare*. The eventual climax is woodland, which may consist of willows if the soil is wet, or other trees such as oak and ash which are more typical of drier inland habitats.

The sequence of stages in the duneland succession (**psammosere**) is summarised in Fig. 9.8.

The sand dune environment

The habitat provided by developing dunes must be one of the most exacting facing any potential plant colonist. Most areas of duneland are characterised by fierce onshore winds which keep the surface sand in a continual state of movement. Stabilisation is achieved by the roots and rhizomes of rapidly growing plants such as marram grass and sedges. Once established, these also exert an important shielding effect, reducing the wind speed at the sand surface by 75 per cent or more. Some typical measurements made by Salisbury [9.2] are summarised in Fig. 9.9.

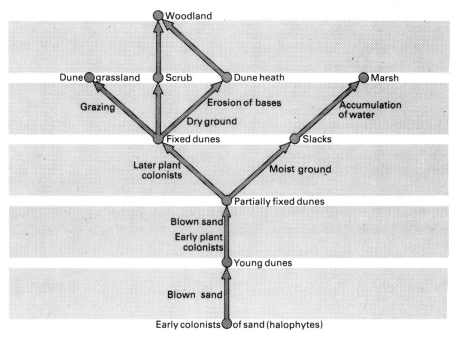

Fig. 9.8 Sequence of duneland succession (psammosere)

The material blown inland from the dry shore consists not only of inert particles of sand but also contains such ingredients as broken pieces of small shells (with a high calcium content) and fine particles of organic matter derived largely from the decomposition of dead seaweed. The dune soil and habitats derived from it are therefore usually alkaline and well supplied with calcium compounds, while the sand surface, once stabilised, often contains an appreciable layer of humus. This explains why mosses such as *Tortula ruraliformis* are often such successful early colonists.

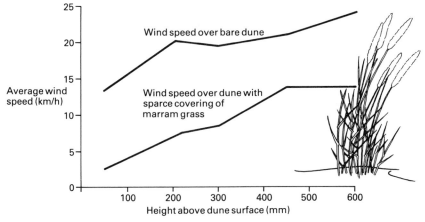

Fig. 9.9 Influence of marram grass, *Ammophila arenaria*, in controlling wind speed over sand dunes (data from Salisbury)

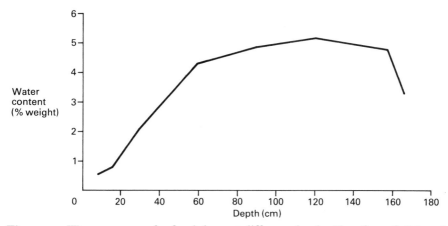

Fig. 9.10 Water content of a fixed dune at different depths (data from Salisbury)

The surface layer of humus is, however, seldom deep and it is therefore subject to periodic wetting by rain followed by rapid drying. The majority of the water retained has penetrated into the layers below (Fig. 9.10). It will be seen that in the dune studied by Salisbury [9.2], moisture attained a maximum at a depth of 120 cm but declined sharply thereafter. The rooting systems of dune-inhabiting plants are well adjusted to such conditions. As on downland, many produce an abundance of shallow roots to catch the periodic rainwater and also a much deeper system to tap the more permanent sources lower down (Fig. 9.11).

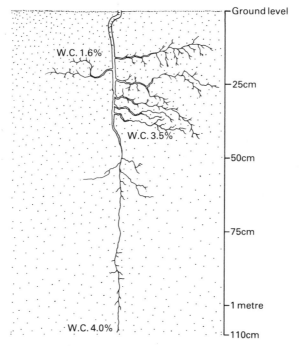

Fig. 9.11 Rooting system of the Portland spurge, *Euphorbia portlandica*, from a partially fixed dune, and the percentage water content by weight (after Salisbury)

Animals in sand dunes

The ever present danger of desiccation exerts a strong influence on the animal colonists of sand dunes. Animals living there frequently bear a close resemblance to the populations found in the splash zone of the sea shore, which is discussed in Chapter 11. They include a variety of flies, spiders, and crustaceans. A few species, by virtue of their peculiar requirements, have managed to overcome the problems posed by the dune environment and are found nowhere else. An example is the little Portland moth whose larva feeds on sea wormwood, *Artemisia maritima*, creeping willow, *Salix repens*, and several other duneland plants.

Behind the dunes the animal communities tend to be typical of other comparable habitats inland. As on downland, the most important mammal is the rabbit. Not only does its browsing contribute materially to the maintenance of dune grassland, but in the vicinity of the dunes themselves large numbers of rabbit burrows can have strong destabilising effects. This is particularly true on the windward side of young and partially fixed dunes where loosened sand can be dispersed by the wind and resulting in a 'blow-out'. To avoid erosion due to human trampling, duck boards are sometimes laid along much frequented pathways (Fig. 9.12).

Some topics for investigation

1 What is the distribution pattern of (a) annuals, (b) perennials on dunes of different ages? What effects do these distributions have on the development of the dune systems?

2 Calculate the diversity index for samples of plants growing on dunes of different ages. To what extent does the index relate to the age of the dune?

Fig. 9.12 Duck-boards laid along a much frequented pathway to reduce erosion (Braunton Nature Reserve)

3 Water is an important factor governing the colonisation of dunes by plants. How does the water content (expressed as percentage weight) of soil at different levels relate to the root systems of (a) annuals (ephemerals), (b) perennials?

4 How does the mode of growth of plants such as marram grass and sedges lead to the rapid colonisation and stabilisation of young dunes?

5 Many plants colonising dunes show strong xeromorphic characters, for example marram grass (rolled leaves with a thick cuticle and stomata sunk in pits) and stonecrop (fleshy leaves with much water tissue). Other common inland plants which are faced with adapting to salt and drought also adopt a xeromorphic appearance, e.g. by becoming more fleshy. In a sample area, what examples of such adaptations occur?

6 To what extent is it possible to trace the stages of succession in a particular dune system? How far has the process advanced towards a climax vegetation (see Fig. 9.8)?

7 The burrowing and grazing of rabbits and human trampling can have significant effects at all stages of dune succession. What evidence is there of this and what have been the outcomes of (a) rabbit activity, (b) human activity?

8 In the vicinity of a 'blow out' what evidence is there of (a) its cause, (b) its effects on the neighbouring dune systems?

Summary

1 Downland soils are usually alkaline and shallow with moisture concentrated near the surface.

2 The climax vegetation on downs is woodland, usually oak. Succession is, however, held in check by the grazing of rabbits and other herbivorous mammals such as sheep.

3 A feature of many downland plants is a rosette pattern of growth, possibly evolved as a response to grazing pressure over a long period.

4 Like its plants, downland possesses a characteristic community of animals. Butterflies and day-flying moths are particularly common. An abundance of calcium in the soil promotes heavy colonisation by molluscs such as snails.

5 Sand dunes represent an unstable ecological environment. Physical influences such as wind pattern and erosion play a large part in determining their eventual size.

6 Plant succession on dunes provides a satisfactory topic for study. In general it passes through three fairly distinct phases which are usually easy to recognise.

7 In areas behind a dune system succession may result in marsh, heath or grassland, depending on circumstances.

8 The sand dune environment is one of the most exacting that can face any potential plant colonist. The continual wind promotes unstable and dry conditions, often with a high salt content. Blown sand containing fragments of shells is responsible for the high calcium content of dune soils.

9 The majority of the animals colonising dunes are arthropods and molluscs. Among mammals, rabbit populations are often large and exert an important control on the vegetation through browsing. Their burrowing can have a destabilising effect on young dunes.

10 Erosion due to human trampling can be a serious problem in sand dunes and necessitates the provision of artificial covering on much frequented pathways.

References

9.1 Brodie, J. (1983) *Grassland Studies*, George Allen & Unwin
9.2 Salisbury, Sir E. (1952) *Downs and Dunes*, Bell
9.3 Dowdeswell, W. H. (4th edn 1975) *The Mechanism of Evolution*, Heinemann
9.4 Nuffield A-Level Biological Science (1970) *Organisms and Populations* (Laboratory Guide), Penguin
9.5 Ranwell, D. (1974) *Saltmarsh and Sand Dunes*, Chapman and Hall
9.6 Hepburn, I. (1952) *Flowers of the Coast*, Collins

10 Ecological communities

Ponds and streams

Compared with the terrestrial habitats described in the previous chapter, waters of all kinds provide a relatively stable environment for plant and animal life. Moreover, the extent and composition of aquatic habitats are usually fairly easy to define and comprehend, and this makes them particularly suitable for the study of ecological principles.

The nature of all aquatic environments depends upon the following factors which exert their effects in varying degree:

(i) the **geological formation** in which they occur. In fresh water this will influence the nature of dissolved substances, while on the sea shore the kind of rock will determine the rate of erosion by the sea and hence the slope and other physical characteristics.

(ii) the **chemical composition** of the water. In the sea this remains remarkably constant, while in estuaries dilution with fresh water leads to a highly variable physical environment. In ponds and streams, water flowing over calcareous rock such as chalk or limestone will tend to be 'hard' with a high pH value. Such conditions favour calcicole plants and animals with calcareous shells such as gastropod molluscs and crayfish.

(iii) the **depth of the water**. This has the important secondary effect of reducing the penetration of light and hence the degree of colonisation by plants.

(iv) the **existence of a current**, which characterises streams and rivers. A current tends to wash away much of the plankton present and hence influences the pathways by which energy can enter the ecosystem.

(v) the **buoyancy of the water**. Since water is a denser medium than air, it (particularly sea water) is more buoyant, and the upthrust exerted on plants and animals enables them to survive without having to support their own weight. This means that their structure can be of a more diverse and delicate kind than that necessary for life on land.

(vi) **environmental changes**. In common with other habitats, all water is subject to the effects of seasonal variations. Diurnal changes such as fluctuations in temperature, can also be considerable, particularly in shallow stretches. The most extreme situations occur on the sea shore and in estuaries, where the effects of the tides can pose problems of desiccation and osmoregulation which influence the distribution of the populations that live there.

Ponds

The word 'pond' is used here in a collective sense to include all areas of static, fresh water such as small pools, canals, reservoirs and lakes. All of these can provide excellent conditions for study, but from the point of accessibility there is much to be said for choosing small areas of water such as those accumulating in disused gravel pits (Fig. 10.1). These are often relatively easy to sample and can

Fig. 10.1 Pond formed from a disused gravel pit with good accessibility for study

house extensive ecosystems of plants and animals. One of the advantages of using a freshwater habitat to illustrate basic ecological principles is that the diversity of species is relatively small compared with those on a sea shore or on land. This greatly eases the problem of identification. Moreover, areas of static water, particularly if they are shallow, are subject to marked climatic variations such as changes in illumination and temperature, which can occur within quite short distances both vertically and horizontally, thus facilitating comparison of one situation with another. Seasonal changes can also exert profound effects on the composition and distribution of populations. Bearing in mind such diversity, the principle stated elsewhere still applies, namely that, to start with at least, it is better to study a small locality intensively rather than attempt to cover a wide area only superficially.

A number of physical factors, sometimes acting singly but more often in combination, exert a predominant effect on the spatial distribution of plant and animal populations in static fresh water.

Light
There are three main ways in which incident light plays an important part in influencing the density and composition of a pond community.

(i) By far its most important effect is on the rate of photosynthesis of plants. Most static waters are fairly turbid and Secchi disc visibility (p. 106) seldom extends below about a metre. However, isolated measurements of illumination at different depths can be misleading, for these will depend on the incident light at that moment. As the graph in Fig. 7.1 (p. 128) shows, over a period of only thirteen days and measured at the same time each day, this can vary by a factor of five or more. Much of the light striking the surface of water is reflected, but the extent varies depending on such factors as its turbidity and the amount of floating vegetation. Again, within the optical spectrum, water absorbs the various wavelengths differentially, the maximum absorption being in the longer wavelengths. Thus the light that penetrates the deeper water tends to be deficient in red and orange light which is the most effective for photosynthesis. The fact that green plant life seldom exists at depths below a metre is thus due to a combination of circumstances, not just to intermittent variations in solar energy falling on the surface.

(ii) Light is essential for animal vision. However, it follows from (i) above that at depths greater than about a metre, the bottom-living fauna must exist in almost total darkness. Bearing in mind that these are largely detritivores, this is no great disadvantage. The active predators which frequent the surface waters are all either insects or vertebrates. Beetles, bugs, and the nymphs of dragonflies all have the compound type of eye which is well adapted for working at fluctuating levels of illumination. Vertebrate carnivores such as fishes and newts have relatively poor eyesight and many of them are colour-blind. This is not a handicap in most freshwater habitats where the surroundings are usually of a sombre and uniform kind and the important requirement is to be able to distinguish tone rather than colour.

(iii) Responses to light in the form of positive and negative taxis have important survival value for pond animals. In general, active predators existing near the surface tend to show positive responses to light while inhabitants of mud such as various mayfly nymphs, crustaceans like the freshwater shrimp, *Gammarus pulex*, and the water louse, *Asellus*, and bivalve molluscs such as the pea-shell cockle, *Pisidium*, tend to shun light. Some of the most sensational responses to light occur among the plankton. Thus, in large lakes such as Windermere, it has been found that water fleas (order Cladocera), notably *Cyclops strenuus*, migrate daily up and down through a distance of about 40 metres in harmony with changes in light intensity. At midnight the maximum concentration is at the surface, while at noon most of the population descends 30 metres or more, the pattern of distribution repeating itself with a cyclical regularity. Precisely what such responses signify is still uncertain, but their occurrence is evidently widespread.

In ponds, where the volume of water is often quite small, the movements of planktonic organisms in response to light are sometimes more difficult to discern. Fig. 10.2 shows some typical results of sampling the *Daphnia*

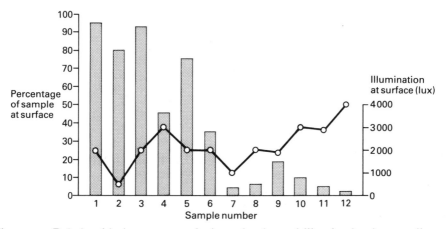

Fig. 10.2 Relationship between zooplankton density and illumination in a small pond

population in a small pond over a period of a few weeks in October. While there is some evidence that a low light intensity may have favoured a concentration at the surface, the situation could have been complicated by other factors such as changes in temperature and short-term fluctuations in population density.

Temperature

Water possesses certain thermal properties which make it a unique environment for life. In particular, it has a high specific heat; 4.2 joules ($= 1$ calorie) are required to raise one gram of water through one degree centigrade. It also has a high latent heat of fusion ($336\,\mathrm{Jg^{-1}}$) and the highest known latent heat of evaporation ($2.25\,\mathrm{kJg^{-1}}$). The combination of these properties explains why temperature variations occurring in water are less than those on land. Moreover, when they occur, they take place far more slowly. Most aquatic organisms are well able to tolerate the normal range of temperature variations that occur throughout the year and are known as **eurythermous**. Species with a narrow temperature tolerance (**stenothermous**) tend to be restricted in their distribution to particular habitats such as the sources of mountain streams, where the temperature of the water changes little throughout the year.

Another important physical feature of water is that it achieves its maximum density at 4°C. Above and below this temperature it expands and therefore becomes lighter. This accounts for the fact that the deeper waters of ponds and lakes never freeze solid in cold weather – an important consideration for the animals that live there. A typical annual record of temperature changes in a loch at a depth of 100 cm is shown in Fig. 10.3.

For many plants and animals in fresh water, the onset of winter and a reduction in temperature necessitates an overwintering phase. Most true aquatic plants are perennials and some such as starwort, *Callitriche stagnalis*, merely sink to the bottom, remaining there in a dormant state until the spring. Numerous species such as the pondweed *Potamogeton*, form underground rhizomes which provide an effective means of vegetative propagation. Many also flower during the summer and form seeds which survive the winter. Hornwort, *Ceratophyllum demersum*, is so well adjusted to life in water that its flowers are pollinated when submerged, the seeds overwintering in the mud. Animals, too, may overwinter in a variety of

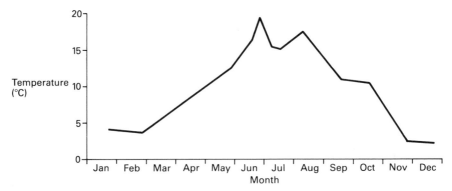

Fig. 10.3 Annual record of temperature in a loch at a depth of 100 cm (after Mills)

phases such as eggs, nymphs, larvae, and pupae, or may hibernate as adults. The community of a pond thus presents a very different picture in winter from that in summer and no overall study is complete without at least one visit during each season.

Oxygen

The surface and shallow waters of ponds and lakes usually contain an abundance of dissolved oxygen and carbon dioxide. Indeed, in areas of submerged vegetation during summer, the concentration of oxygen often achieves supersaturation. In the mud below, the level rapidly declines and may assume the anaerobic conditions characteristic of bogs (p. 243), as is shown in Table 10.1.

Table 10.1 Oxygen concentration at different levels in a ditch (March)

Level	% saturation with oxygen
Water 15 cm below surface	63.5
Water 65 cm below surface (above mud)	41.2
Mud 15 cm below surface of mud	33.3
Mud 30 cm below surface of mud	25.8
Mud 45 cm below surface of mud	29.6

The figures above were obtained by the porous pot and Winkler method (p. 115) and illustrate that in the absence of photosynthesis during winter the oxygen concentration in the surface water drops considerably. This is of little consequence to the animals living there as their metabolic rate will also be reduced by the lower temperature. However, oxygen shortage is a permanent feature of the anaerobic conditions in deep mud and this poses respiration problems for animal colonists.

The availability of oxygen and problems of gas exchange are matters that mainly concern the animal community and a variety of adaptations exist which help to overcome these problems (Fig. 10.4).

(i) **Air obtained through the body surface**
Small animals such as waterfleas (Crustacea), which make up the zooplankton have a large surface area in relation to their bulk and frequently possess no special respiratory organs. In these, the whole body acts as a gill. In response to a reduced oxygen supply, as may occur during winter (see Table 10.1), some of them, for example *Daphnia*, are able to synthesise haemoglobin. So the pink colour of the animal serves as a rough guide to the oxygen content of the water.

(ii) **Respiratory pigments**
Reference has already been made to the temporary production of haemoglobin under conditions of reduced oxygen content. Haemoglobin is also a permanent feature of some burrowers in mud that have to withstand continuous anaerobic conditions. These include annelids such as the square-tailed worm, *Eiseniella*, and the tube-making *Tubifex*, also the larvae of some midges of the genus *Chironomus* (sometimes confusingly referred to as 'blood worms').

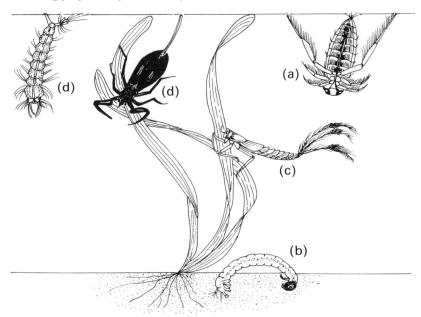

Fig. 10.4   Some methods of gas exchange in freshwater animals: (a) transporting an air bubble from the surface – water boatman, *Notonecta glauca*; (b) blood gills – larva of midge, *Chironomus*; (c) tracheal gills – nymph of dragonfly, *Chloeon dipterum*; (d) breathing siphon – larva of gnat, *Culex pipiens*, and water scorpion, *Nepa cinerea*

(iii) **Gills**

A gill is a respiratory structure which absorbs and releases gases in solution, the moisture being provided by the environment. In aquatic animals gills are of two kinds:

(a) blood gills – projections of the body into which the blood system flows and which are in direct contact with the surrounding water. These are not only characteristic of fishes but are found in other small organisms such as the larvae of some midges and caddis flies.

(b) tracheal gills – thin, plate-like structures permeated by branches of the tracheal system. They are found only in arthropods, mainly in insects. They are particularly well illustrated in the nymphs of certain dragonflies such as *Chloeon dipterum*, also in the nymphs of mayflies and alderflies where variations in their structure can provide useful features for identification.

(iv) **Atmospheric air through spiracles**

The lack of a current in pools means that animals relying on the atmosphere for their oxygen supply can rise to the surface to obtain fresh supplies through apertures (spiracles) without the danger of being washed away. The mechanisms involved are of several different kinds:

(a) breathing tubes and siphons   Many insects such as the larvae of the gnat, *Culex pipiens*, and the water scorpion, *Nepa cinerea*, have well-developed siphons which enable them to pierce the surface film (Fig. 10.5). A few bottom-living forms in shallow water also use this method of gas exchange, for example, the larvae (sometimes known as rat-tailed maggots) of

Fig. 10.5 Water scorpion, *Nepa cinerea* × 2. Note the breathing siphon piercing the surface film (photograph S. Beaufoy)

the drone-fly, *Eristalis*, where the siphon is telescopic and can extend to a length of 6 cm or more. Siphon-breathers face the danger of waterlogging, and their siphons invariably have some means of forming an air lock to prevent the entry of water. This is usually in the form of a set of fringing hairs which entrap air bubbles.

  (b) transportation of atmospheric air A number of species such as beetles and water bugs, surface periodically to obtain a new supply of atmospheric air. The air may be carried as bubbles located close to the openings of the tracheal system (spiracles). For example, in the water boatman, *Notonecta glauca*, a single bubble is trapped among hairs at the posterior of the abdomen. Alternatively, the air may form a sheen-like layer covering the whole body, as in the bug, *Corixa*.

  (c) swallowing air Among fishes, members of the carp family (Cyprinidae) are capable of swallowing air and absorbing oxygen through the wall of the intestine, which is well supplied with blood vessels, and therefore acts as a kind of lung. Small species such as the minnow, *Phoxinus laevis*, can be observed in an aquarium to rise to the surface periodically to gulp a mouthful of air. This explains why the goldfish, which belongs to the same family, is able to tolerate conditions of domestication that would be lethal to any other aquatic animal.

(v) **Lungs**

In contrast to gills, lungs are respiratory structures kept moist by the organism. It follows that they are always located in some sort of branchial chamber. Apart from mammals, true lungs are rare among aquatic animals and are confined to gastropod molluscs (snails). For the purpose of classification, the gastropods are divided into two groups according to their method of gas exchange:

(a) those with an **operculum** – a horny plate on the foot which closes the aperture of the shell when the animal has contracted. There are ten freshwater species in Britain, all of which breathe by gills and are largely, but by no means always, confined to running water where the level of oxygenation is high. An example is Jenkins's spire snail, *Hydrobia jenkinsi*.

(b) those without an operculum in which the inner surface of the mantle serves as a lung (**pulmonates**). This group includes all the common pond snails such as the genera *Limnaea* and *Planorbis*.

The contrasting behaviour of the two groups is seen when they are kept in an aquarium. Jenkins's spire snail can remain submerged indefinitely, while pulmonates such as the wandering snail, *Limnaea pereger*, have to surface periodically to breathe.

pH and dissolved minerals

The precise influence on aquatic plants and animals of variations in the concentration of hydrogen ions is difficult to determine since the presence or absence of these ions is almost invariably governed by other ions. Nonetheless, the pH of an environment can change rapidly over a short distance and this is often associated with marked variations in population distribution. An example is provided by a muddy ditch where oxygen concentrations were estimated at different depths (Table 10.1 p. 185). Corresponding estimates of pH made at the same time are summarised in Table 10.2.

Table 10.2   pH at different levels in a ditch (March). See also Table 10.1 (p. 185)

Level	pH
Water 15 cm below surface	8.5
Water 65 cm below surface (above mud)	7.5
Mud 15 cm below surface of mud	7.0
Mud 30 cm below surface of mud	6.5
Mud 45 cm below surface of mud	6.5

Evidently, a reduction in oxygen (Table 10.1) corresponded to a fall in pH, although how the two were related is uncertain. Sampling the animal populations in the mud revealed a variety of midge larvae, particularly those of *Tanypus* and *Chironomus*. While these were abundant in 15 cm of mud, they were almost absent at 30 cm.

In ponds and lakes the range of pH is about 8.5–4.7. Animals feeding on detritus and therefore only indirectly dependent on green plants for their food, seem to be able to tolerate a wide range of pH without ill effect. Thus the freshwater shrimp, *Gammarus*, appears to survive equally well in habitats covering the whole range of pH quoted above.

As we have seen already (p. 166) the distribution of plants is greatly affected by pH changes in the soil water, which are associated with the presence or absence of calcium ions (Ca^{2+}). Indeed, we can divide plants for convenience into calcicole (lime tolerators) and calcifuge (lime haters). This classification applies to freshwater species as well as those on land. It therefore follows that where calcium (Ca^{2+}) ions are present and the pH is high, the community of aquatic animals which feed on the plants will be somewhat different from when calcium ions are absent.

Besides influencing the plants which constitute their food, calcium can have an important direct influence on the distribution of animal populations. For instance, it is an important component of the shells and integuments of molluscs and crustaceans; it also has a significant effect on the permeability of membranes. This may well account for the fact that among planarian flatworms the common *Polycelis nigra* occurs in water at all values of pH while the milky-white *Dendrocoelum lacteum* is only found where there is much calcium and the pH is high.

An interesting example of the effects of calcium ions is quoted by Macan [10.1] and concerns populations of the water louse, *Asellus* (Fig. 10.6) sampled in tarns in the Lake District and elsewhere. Two sets of data were obtained by Moon and Reynoldson and these are summarised in Fig. 2.3 (p. 17). Although the results

Fig. 10.6 Water louse, *Asellus* × 3. Its distribution is influenced by the presence of calcium ions (photograph John Clegg)

are somewhat conflicting, it nonetheless seems clear that *Asellus* was virtually absent from waters with a calcium content of less than 10 mg l^{-1} but nearly always present when the value exceeded 12.5 mg l^{-1}.

Variation in dissolved minerals

In the previous account we have made the implicit assumption that the concentration of minerals in water, and hence its pH, remains constant. However, in most ponds which contain a fair number of green plants, this is not so. From March onwards, the temperature of the water rises, the concentration of dissolved salts such as nitrate reaching a maximum as a result of the breakdown of detritus by aerobic bacteria during the winter. Conditions are thus favourable for the great outburst of plant plankton (algae and diatoms) which is a feature of static waters in spring (Fig. 10.7). Predation by small herbivores and a depletion of minerals contribute to a marked reduction of the phytoplankton in July to September. During the warm spells in late September and October the algae sometimes stage a partial recovery (Fig. 10.7) as the supply of mineral salts starts to increase once more. The pattern of algal growth thus conforms closely to that of temperature changes in the surface waters (compare Figs 10.7 and 10.3 p. 184).

Pond communities

In most ponds it is possbile to discern three main zones of life:

(i) *The surface film*
This is the interface between air and water and a region of abundant oxygen and maximum illumination.

The commonest plant colonists are the duckweeds, particularly *Lemna minor* (Fig. 10.8) with their short trailing roots hanging down in the water. One of the most striking surface colonists is the water fern, *Azolla filiculoides*, a North American plant which has escaped from aquaria and become naturalised in many places. In the autumn its fronds turn red (Fig. 10.9).

The animals of the surface film are a specialised group of air-breathing insects which can move at considerable speed. Their weight is supported by surface tension on their legs, as can easily be demonstrated in an aquarium by adding a

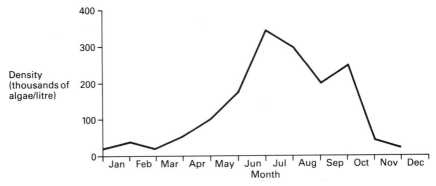

Fig. 10.7 Annual variation in the density of plant plankton in a pond

drop of detergent to the water, when they immediately sink. All the larger species such as the water cricket, *Velia currens,* and pond-skater, *Gerris* sp (Fig. 10.10), are carnivores preying on small animals visiting the surface from below or those that drop into the water from above. Some of the smaller animals like the springtails (Collembola) are capable of jumping on the surface film as if they were on land. On the lower side of the surface there are no permanent colonists, but, as described already, many species visit it intermittently in quest of atmospheric air.

(ii) Vegetation zone
This extends from the margins of a pond to the middle and includes three different categories of plants all rooted in the mud. At the edge there are those that are near to becoming land plants which have their leaves and stems in air, such as the large bulrush, *Typha latifolia,* and the smaller fool's watercress, *Apium nodiflorum,* and watermint, *Mentha aquatica.* In mid-water are species with submerged stems and floating leaves such as the water lily, *Nuphar lutea.* The

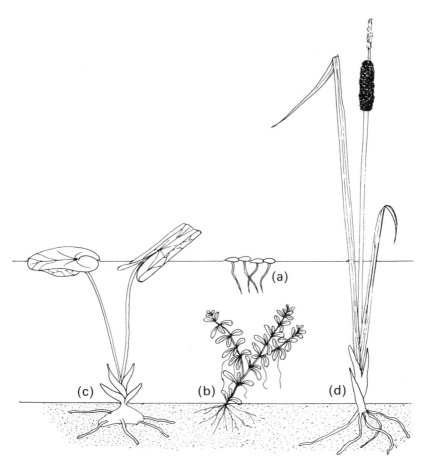

Fig. 10.8 Some typical plants of the surface and vegetation zones of a pond: (a) free-floating – duck weed, *Lemna minor*; (b) fully aquatic with roots in mud and all parts submerged – Canadian pondweed, *Elodea canadensis*; (c) leaves floating, roots in mud – water lily, *Nuphar lutea*; (d) leaves aerial, roots in mud – bulrush, *Typha latifolia*

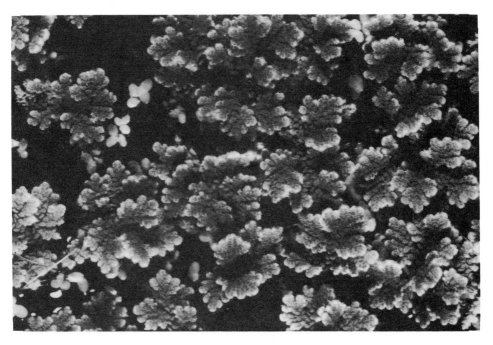

Fig. 10.9 Water fern, *Azolla filiculoides*, an aquarium escape which colonises the surface of ponds (photograph John Clegg)

majority, however, are totally submerged and often form dense masses just below the surface (Fig. 10.8). They include the water crowfoot, *Ranunculus fluitans*, the hornwort, *Ceratophyllum demersum*, and Canadian pondweed, *Elodea canadensis*. The water starwort, *Callitriche palustris*, is peculiar as it has both floating and submerged leaves and is able to colonise both relatively static and flowing water successfully (Fig. 10.18 p. 203).

The vegetation zone is a region of high productivity and supports a wide range of herbivorous animals, which in turn provide food for numerous carnivorous species, both small and large. Some of the ecological niches to be found in a pond and their relationships are summarised in Fig. 10.11. Typical animals are included in Fig. 10.10.

The open water is, above all, the home of a vast number of minute floating plants and animals (plankton). The plants consist mainly of two groups, the algae (Fig. 10.12(a)) and the diatoms (Fig. 10.12(b)). They provide the food for an abundant animal plankton (Fig. 10.13). Some food relationships in static fresh water are shown in Fig. 10.11. At the other end of the size range are the fishes such as the minnow, *Phoxinus laevis*, and three-spined stickleback, *Gasterosteus aculeatus*, both of which are mainly carnivorous. However, several common species such as the roach, *Rutilis rutilis*, eat a good deal of plant food as well.

(iii) Mud

As we have seen already, mud is a zone with a high organic content in the form of detritus, and serves two important functions. It provides a source of nutrition for

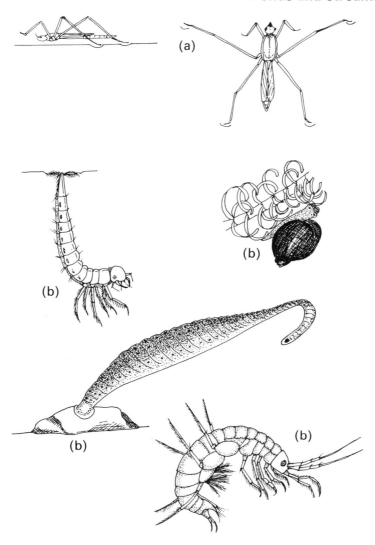

Fig. 10.10 Some typical animals of the surface and vegetation zones of a pond:
(a) surface film – pond skater, *Gerris*; (b) vegetation zone – carnivorous larva of great
diving beetle, *Dytiscus marginalis*, leech, *Glossisiphonia* (parasite), pond snail, *Limnaea*
(herbivore), and freshwater shrimp, *Gammarus pulex*, (detritivore)

aerobic microorganisms, which break the detritus down to mineral salts and
restore the fertility of the water above. It also serves as a home for numerous
burrowing species like the annelid worm, *Tubifex*, midge larvae such as
Chironomus, and a number of filter-feeding bivalve molluscs ranging from the tiny
pea-shell cockle, *Pisidium*, which attains a length of around 5 mm, to the large
swan mussel, *Anodonta cygnea*, which can grow to 15 cm or more. Mud, together
with its detritus, thus provides an important ecological habitat for part of a pond
community. Some of its food relationships are shown in Fig. 10.11.

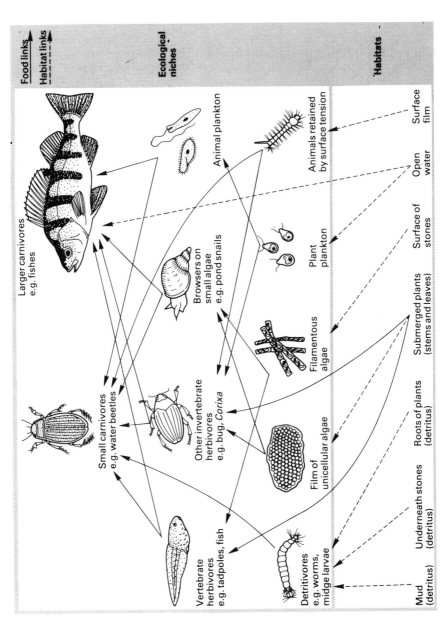

Fig. 10.11 Some habitats and ecological niches in a pond

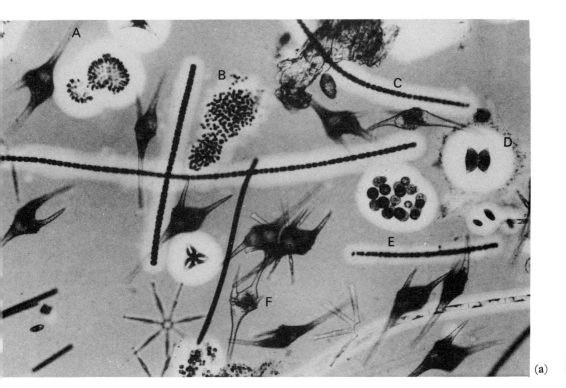

(a)

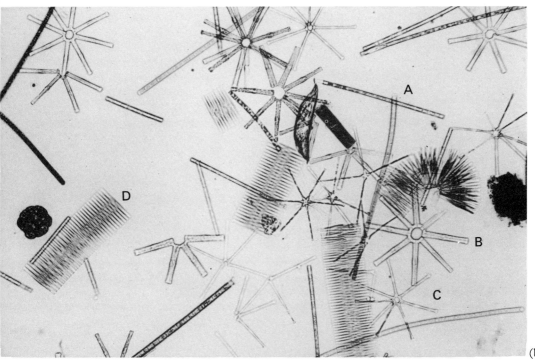

(b)

Fig. 10.12 Phytoplankton in fresh water. (a) Algae: A – *Gomphosphaeria*;
B – *Microcystis*; C – *Anabaena*; D – *Staurodesmus* (desmid); E – *Eudorina*; F – *Ceratium*
× 160. (b) Diatoms: A – *Melosira*; B – *Tabellaria*; C – *Asterionella*; D – *Fragilaria* × 160
(photograph Hilda Canter Lund F.R.P.S.)

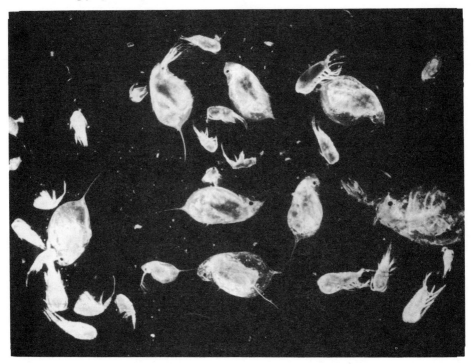

Fig. 10.13 Zooplankton in fresh water, mostly small crustaceans such as daphnids and copepods (photograph John Clegg)

Plant succession in static fresh water

The gradual accumulation of mud which tends to take place round the edges of ponds and lakes results in a characteristic successional sequence (**hydrarch succession** or **hydrosere**) from water to land. Its principal features are a gradual reduction in the depth of the water, the raising of the soil above the water level and the eventual lowering of the water table. The rate of the process and the outcomes of the different stages vary somewhat with circumstances but a simplified picture of the early stages is shown as a profile chart in Fig. 10.14. As the level of the mud rises, submerged plants are replaced by those with floating leaves and these in turn give way to a swamp community characterised by such plants as the bulrush, *Typha latifolia*, with aerial leaves and stems but rooted in the water. In acid conditions a bog will result, but as the level of the ground rises and conditions become drier, this will give rise to heathland (see Chapter 12). In basic soils with a pH of around 6.5 or more, a further raising of the soil level will enable land plants to colonise the bank. In wet conditions these will include species typical of water meadows and fens such as the great water grass, *Glyceria maxima*, and the great water dock, *Rumex hydrolapathum*. Dry conditions will lead to the formation of the different types of grassland described in Chapter 9. A simplified sequence of these successional stages is summarised in Fig. 10.15.

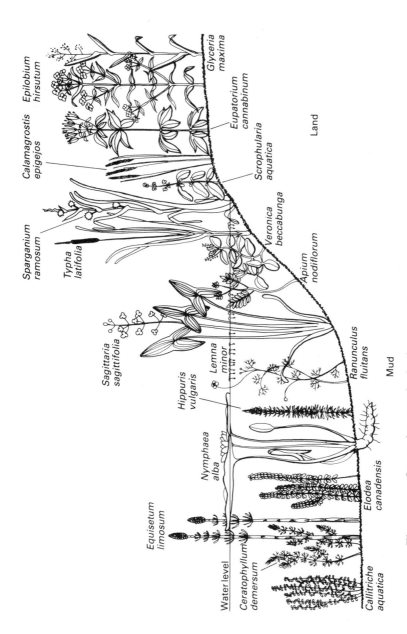

Fig. 10.14 Some plant species in the successional sequence (hydrosere) in a pond

Epilobium hirsutum

Calamagrostis epigejos

Glyceria maxima

Eupatorium cannabinum

Scrophularia aquatica

Sparganium ramosum

Typha latifolia

Land

Veronica beccabunga

Apium nodiflorum

Sagittaria sagittifolia

Lemna minor

Hippuris vulgaris

Ranunculus fluitans

Mud

Nymphaea alba

Equisetum limosum

Water level

Ceratophyllum demersum

Elodea canadensis

Callitriche aquatica

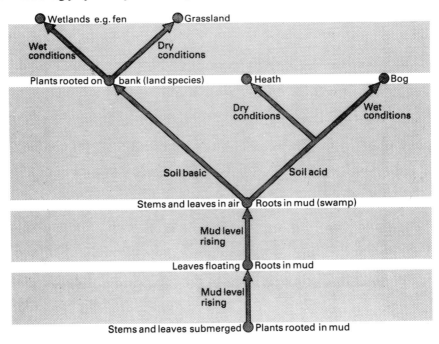

Fig. 10.15   Simplified sequence of plant succession (hydrosere) from static fresh water to land

Some topics for investigation

1   In a small part of a pond, identify the various microhabitats present (Fig. 10.11). Using appropriate sampling methods (p. 73) record (a) the plant and animal colonists of each habitat, (b) their relative numbers (counting is laborious and it will probably be sufficient to classify the populations arbitrarily as, say, very common, common, rare). Judged by their distribution, which species (a) are most adaptable, (b) have the most specialised requirements? If possible, compare two areas, for example, one shaded and the other in the sun. Can you account for any differences?

2   From 1 above, which of the animals are carnivores and which herbivores, judged by their distribution, behaviour, and mouthparts? In any aquatic habitat (a) identify a few links in the food web, (b) draw up a pyramid of numbers (p. 54) for the organisms present. What are the limitations of this method of recording the flow of commodities in an ecosystem? How could it be improved and what alterations would this necessitate in the techniques of study?

3   To what extent does plankton density (plant and animal combined) vary with (a) depth, (b) time of day? Sampling could consist of standardised 'unit sweeps' with a plankton net over a fixed distance. Numbers can be counted and individuals identified in a cavity chamber (p. 93). Estimate the standard deviation of the means of samples and the standard error to compare contrasting batches (p. 131). Are any observed differences significant? If so, how do you account for them?

4 Using a resistance thermometer (p. 104) and light meter (p. 106), estimate the gradient of temperature (thermocline) and light at different depths. Is the distribution of plankton related to either, both, or neither of these variables? If possible, repeat the sampling at night. How does the distribution of plankton differ from that in daytime?

5 Is there any evidence of a plant succession from water to land (**hydrosere**) in the pond and its surroundings. If there is, determine the stages through which succession may be taking place. What are the principal factors (a) promoting, (b) inhibiting succession?

6 Mud is a difficult medium to sample but it is usually possible at the edge of a small pond using a shovel, trowel or auger (p. 99). How does the mud-dwelling animal community (sampled at different depths if possible) compare with the vegetation community in terms of (a) density and variety of species, (b) mode of life including feeding methods?

7 With an oxygen electrode (p. 115) (or using the Winkler technique), measure the oxygen concentration at different levels, including the mud. How do the breathing mechanisms used by the animals at different depths relate to the differing oxygen availability?

8 From a sample of animals from the vegetation zone, calculate an index of diversity (p. 22) and compare it with the situation elsewhere, e.g. sun/shade or pond/stream. How do you account for any differences observed?

9 Set up a miniature ecosystem in an aquarium. (It must be well aerated.) Include organisms such as the water plants *Elodea* and *Callitriche*; snails, e.g. *Limnaea* and *Planorbis*; caddis larvae, e.g. *Agapetus*; crustaceans, e.g. *Gammarus*; bugs, e.g. *Corixa* and *Notonecta*; and fishes, e.g. minnows or sticklebacks (the number of carnivores should be kept to a minimum). How does the behaviour of the different species vary in terms of (a) feeding, (b) breathing?

10 Set up a Margalef-type laboratory model simulating succession in pond water plankton (p. 69). How is the rate of succession influenced by (a) light intensity, (b) temperature? What is the successional climax?

Streams

The term 'stream' is used here collectively to cover any kind of water that is flowing, and thus ranges from slow rivers to torrential brooks. Two extremes are contrasted in Figs. 10.16 and 10.17 and both have good accessibility for study. Although all rivers have many plant and animal species in common, some deviate into small backwaters where the water is virtually still and these house miniature pond communities, particularly of surface-dwelling insects.

Many of the features of static water communities discussed in the previous section apply to streams as well. However, the fact that the water is constantly moving is bound to have an equalising effect on the substances dissolved in it, so that marked contrasts in such factors as pH, minerals, and oxygen seldom occur in the different zones of a stream. Exceptions arise from interference by man, such as the release of effluents from industry, or the seepage of fertilisers dissolved in rain which gives rise to an excess of minerals (**eutrophic** conditions). We will return to these and other aspects of pollution in Chapter 14. Although the chemical aspects of the stream environment are relatively stable, its physical nature can fluctuate

Fig. 10.16 A slowly-flowing stream (River Cray) with good accessibility for study

Fig. 10.17 A rapidly-flowing stream (Badgworthy Water, Exmoor). The vegetation indicates base-free conditions

violently. Thus, only a few hours of heavy rainfall will transform a gentle stream into a raging torrent, increasing its depth and rate of flow by a factor of three or more.

Temperature

Streams and rivers often vary considerably in temperature along their length. In general, the nearer their source, the colder they are; this is particularly true of mountain streams. Where they arise, a number of stenothermous (with narrow temperture tolerance) species are to be found, such as the flatworm *Crenobia alpina*, thought to be an ice-age relic. Its optimum temperature for maximum rate of growth has been shown to be 10°C, whereas that of *Dugesia gonocephala*, a eurythermous (wide temperature tolerance) species occurring in warmer waters is about 17°C. The distributional pattern in rivers of the larvae of several different species of caddis flies is also closely related to temperature. Varley [10.2] has provided data on the temperature limits characteristic of three categories of freshwater fish (Table 10.3).

Table 10.3 Temperature limits characteristic of three categories of freshwater fish (after Varley)

Category	Upper lethal limit (°C)	Optimum for growth (°C)	Typical for spawning (°C)
Cold water stenotherms (e.g. trout, grayling)	<28	7–17	<10
Intermediate (e.g. pike, perch)	28–34	14–23	>10
Eurytherms (e.g. carp, tench)	>34	20–28	>15

At the upper limit, a temperature which is lethal for stenothermous species such as the rainbow trout, *Salmo gairdneri*, is within the optimum range for growth of eurythermous forms such as the carp, *Cyprinus carpio*. This helps to explain the different distribution of the two species, the trout being confined to shallow, rapidly flowing water while the carp is found in deeper, slow-flowing rivers and lakes. It should be added that availability of oxygen may also be a limiting factor influencing the distribution of the two species, for while the trout is an active fish with a high metabolic rate, the carp is more sluggish. Fast-flowing water, because of its turbulence, will tend to be saturated with air, or nearly so, in contrast to static water where the oxygen supply may become depleted, particularly during winter. It is significant that the carp and tench both belong to the family Cyprinidae, a group of fish characterised by the ability to supplement their limited oxygen supply by rising to the surface and gulping air, which is then absorbed through a highly vascularised gut wall (see also p. 187).

Current

It is not easy to generalise about the effects of a current on the characteristics of stream communities since conditions tend to vary greatly in different places. However, the following features are common to all flowing waters to some degree.

(i) The eroding impact of a current tends to wash away the smaller and lighter soil particles such as clay, also the finely divided organic matter (detritus). These materials tend to accumulate on obstructions such as stones and the roots of plants, and if not quickly removed will form mud banks and small islands.

When the flow of the water is slow, mud tends to accumulate at the sides of a stream (Fig. 10.16) and supports a characteristic flora and burrowing community of animals.

(ii) The churning action of the water maintains it in a state of almost permanent saturation with air. Moreover, it tends to bring about local uniformity of pH and temperature, and the distribution of dissolved substances. But, as we have seen already, there can be a considerable variation in the physical environment along the length of a water course, with important implications for the distribution of populations.

(iii) The force of the current also tends to wash away all living organisms other than those which are either rooted, capable of finding shelter, or powerful swimmers. Plankton is therefore largely absent from streams and rivers, which accounts for the limited range of species found in them compared with that of still water. It is often surprising that weakly-swimming animals inhabiting the bed of a stream, such as the freshwater shrimp, *Gammarus pulex*, are able to move about at all. But it must be remembered that current speed decreases with depth until, on the stream bed, there is (theoretically at least) no movement of water at all. This fact can be readily demonstrated with the current meter described earlier (p. 110).

Adaptations for survival in a current
In spite of the eroding effects of flowing water, streams frequently support an abundance of plant and animal life which displays a variety of adaptations. Some of these are shown in Figs 10.18 and 10.19, and are as follows:

(i) *Attachment mechanisms* Among the plants in streams, many, such as the water crowfoot, *Ranunculus fluitans*, have well developed rooting systems anchored firmly in the substratum of silt or gravel. The colonists of the surfaces of stones tend to be filamentous algae which have an adhesive outer surface, or mosses such as the willow moss, *Fontinalis antipyretica*, which have well developed rhizoids that are able to penetrate small cracks in the rock surface. Many animals also have adhesive surfaces, ranging from sponges such as *Ephydatia fluviatilis* and small protozoans, to large insects such as the encased larvae and pupae of many caddis fly species.

(ii) *Suckers and hooks* Suckers are characteristic of many small insects such as the larvae of the midge *Simulium*. They are also a feature of leeches and the brook lamprey, *Lampetra planeri*, with its circular mouth. Some caddis larvae such as *Polycentropus*, are web-spinners and these have hooks at the rear of the body which serve as anchors.

(iii) *Body surface* Many stream plants such as the water crowfoot, *Ranunculus fluitans*, have mucilage on the surface of their leaves and are slimy to touch. No doubt this plays some part in reducing friction against the substratum. Among animals, planarians such as the common *Polycelis nigra*, have a slimy adhesive surface, while gastropod molluscs like the wandering snail, *Limnaea pereger*, have a muscular foot which is well supplied with mucous glands.

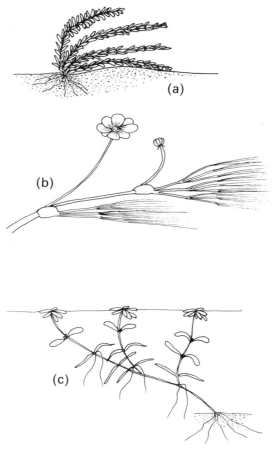

Fig. 10.18 Adaptations among plants for withstanding a current: (a) attachment mechanisms – rhizoids of the willow moss, *Fontinalis antipyretica*; (b) streamlining – water crowfoot, *Ranunculus fluitans*; (c) two kinds of leaves (broad floating and thin submerged) – starwort, *Callitriche*

(iv) *Streamlining* The stems and leaves of many aquatic plants are stream-lined to some extent (Fig. 10.18), while in a few such as the starwort, *Callitriche*, floating and submerged leaves are produced, enabling the plant to flourish in both flowing and static water. The shape of most of the animals found in streams tends to be bluntly rounded in front tapering to a narrow point behind. This is the shape that reduces the 'drag' of the water to a minimum, thus lessening the effect of the current and facilitating more rapid movement. Many animals living in mid-stream have achieved some degree of flattening as well, notable among these being some of the large mayfly nymphs such as *Ecdyonurus venosus* (Fig. 10.19).

(v) *Behavioural responses* In general, stream animals exhibit **positive rheotaxis**, that is to say, they tend to face and move upstream. This facilitates the use of gills in gas exchange and also the capture of food floating towards them. Another response is **positive thigmotaxis** – a tendency to cling to any surface with which the body comes into contact. Such a response

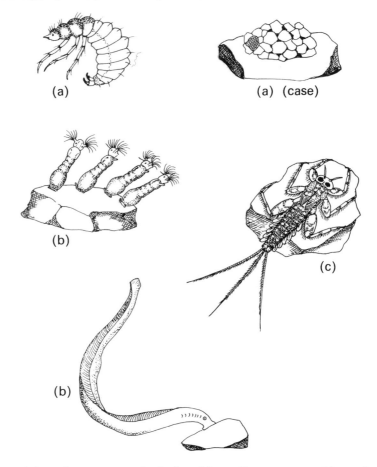

Fig. 10.19 Adaptations among animals for withstanding a current: (a) attachment to stones – larva of caddis fly, *Agapetus*; (b) hooks and suckers – larva of the midge, *Simulium* (hooks) and brook lamprey, *Lampetra planeri* (sucker); (c) streamlining – nymph of mayfly, *Ecdyonurus venosus*, which adopts a crouching attitude on the stone surface

is important for small molluscs such as Jenkins's spire snail, *Hydrobia jenkinsi*, whose powers of locomotion and avoidance of an excessive current are limited.

Behaviour patterns of the kind described here achieve a good deal more than merely enabling animals to maintain themselves in a particular microhabitat. Natural conditions are subject to constant disturbance, not least by the activities of man, with the result that species such as small arthropods frequently find themselves being washed downstream. An instinctive response which stimulates them to climb back to a more favourable situation can thus be of great survival value. As in ponds (p. 183), many stream animals change their distribution at night, moving on to the surface of stones and vegetation. Although the significance of such behaviour is far from clear it no doubt serves an important purpose, for instance in facilitating feeding.

Stream communities

In many ways stream communities of plants and animals, and the factors influencing their distribution, resemble those of ponds. However, there are important differences, and it is these that will be emphasised in this section.

(i) Surface film
The existence of a current prevents permanent colonisation of the surface in streams. However, eddy pools of varying size are a feature of the edges of most rivers and these sometimes support a community of floating plants such as the duckweed, *Lemna minor*, and pond-skating insects. The situation is an unstable one, for during periods of heavy rainfall when the level of the water rises and the current increases, these temporary communities are soon washed away.

(ii) Vegetation zone
The productivity of streams is usually much lower than that of ponds. Nonetheless, a dense growth of weed is sometimes present, consisting of plants rooted in the gravel and sand of the stream bed. Typical species are the water crowfoot, *Ranunculus fluitans*, the water milfoil, *Myriophyllum spicatum*, and Canadian pondweed, *Elodea canadensis*. Their roots, stems, and leaves provide important habitats and food for a variety of animal species some of whose ecological niches and relationships are summarised in Fig. 10.20. The surface of stones is frequently colonised by mosses such as the willow moss, *Fontinalis antipyretica*, but these afford insufficient shelter for most animals.

(iii) Open water
The presence of a current means that only the most powerful swimmers can maintain themselves in the open water. This is colonised largely by fishes. Some species are mainly bottom living such as the bull-head, *Cottus gobio*, which is flattened dorso-ventrally and can wedge itself under stones and among weed, and the stone loach, *Cobitis barbatula*, which has sensory tentacles at the front of its head.

(iv) Gravel, sand, and mud
In spite of the current, much of the detritus resulting from the death and decomposition of stream organisms accumulates under stones, among the roots of plants, and in the mud at the edge. Many streams are overhung in places by trees and shrubs (Fig. 10.21) and leaf-fall in the autumn provides an important input of energy, adding greatly to the accumulation of detritus. A feature of most streams is thus the high density of detritivores. Some of these have characteristic distributions and variations in behaviour pattern. Thus the nymphs of the mayfly, *Ephemerella ignita*, are active and tend to congregate under stones in mid-stream, while those of *Ephemera danica* are burrowing forms and confined largely to the mud at the edge.

Plant succession in streams

The sequence of a hydrosere in flowing water habitats follows much the same pattern as in ponds (Fig. 10.15) but it can take place, at least part way, far more quickly. The natural tendency in streams is for silt to accumulate in the vicinity of

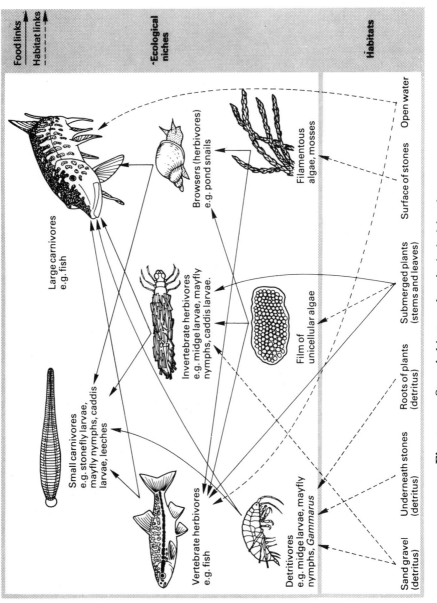

Food links ↑↑

Habitat links ↑↑ (dashed)

Ecological niches

Habitats

Large carnivores
e.g. fish

Small carnivores
e.g. stonefly larvae,
mayfly nymphs, caddis
larvae, leeches

Vertebrate herbivores
e.g. fish

Browsers (herbivores)
e.g. pond snails

Invertebrate herbivores
e.g. midge larvae, mayfly
nymphs, caddis larvae.

Detritivores
e.g. midge larvae, mayfly
nymphs, *Gammarus*

Filamentous
algae, mosses

Film of
unicellular algae

Open water

Surface of stones

Submerged plants
(stems and leaves)

Roots of plants
(detritus)

Underneath stones
(detritus)

Sand gravel
(detritus)

Fig. 10.20 Some habitats and ecological niches in a stream

Fig. 10.21 A stream shaded by trees and shrubs. Leaf-fall in autumn provides an important input of energy

boulders and the roots of plants. One of the principal problems in conservation is to prevent these accumulations and to keep the water flowing. Once it is allowed to build up an island quickly forms, and in the course of a year it can become colonised by a flourishing community of land plants. Such events often occur in streams and are well worth study as examples of hydrarch succession in miniature.

Some topics for investigation

1   Repeat topics 1 and 2 for ponds (p. 198) in a stream, referring to Fig. 10.20 (habitats and ecological niches in a stream). How do the findings from the two ecosystems differ? To what extent can you account for these differences?
2   With an oxygen electrode (p. 115) (or using the Winkler method), compare the oxygen concentration of the water in the vegetation and mud zones of a stream. How are the breathing mechanisms of the animals related to these habitats?
3   An approximate idea of the standing crop (p. 55) (estimated by biomass) can be obtained by removing a few square metres of vegetation together with all the animals present. (The area chosen should be in full sun and not overhung by trees.) Weigh the vegetation after removing as much water and mud as possible, also the herbivores and carnivores. What are the ratios of the trophic levels represented? If the density of fishes can be judged approximately, an estimate of the proportionate biomass of the tertiary consumers could also be made. (*N.B.* Return as much of the living material as possible to the stream after the investigation.)

4   Using an appropriate sampling technique, investigate the distribution of the freshwater shrimp, *Gammarus pulex*, in a small section of the stream. Is there any evidence that the size distribution of the population is related to the rate of flow of the current? If there is, can you account for any differences observed? How would you test your hypothesis?

5   Many other animals have a characteristic distribution in relation to the various microhabitats available, examples being mayfly nymphs (*Ephemerella*) and Jenkins's spire snail, *Hydrobia jenkinsi*. How do their distributions differ both qualitatively and quantitatively? What factors influence their distribution?

6   To what extent is the colonisation of stones by plants such as willow moss, *Fontinalis antipyretica*, related to the rate of flow of the current? A measure of colonisation could be percentage coverage. How do you account for any variation observed between the centre of the stream and the sides?

7   In a stream, what evidence is there of a hydrosere (a) in the formation of a bank or island in mid-stream, (b) at the edge? How does the sequence of events compare with that of a pond (Fig. 10.15)?

8   From samples of stream animals, calculate the index of diversity (p. 22). Compare the situation in mid-stream with that at the side, also with the vegetation zone of a pond (see item 8 p. 199). How do you account for any differences?

9   In a section of a stream, how does the daytime distribution of small animals such as arthropods differ from that at night?

Summary

1   The nature of an aquatic environment is determined by such factors as the geological formation where it occurs, the chemical composition and depth of the water, its buoyancy (salinity), and the extent of environmental changes.

2   The environment of ponds and other areas of static water is determined by a number of physical factors working singly or in combination. These include light, temperature, oxygen, pH, and dissolved minerals.

3   In most pond communities it is possible to discern four main zones of life: the surface film, the vegetation zone, open water, and mud.

4   Plant succession in static water (hydrosere) follows a predictable sequence of stages. The end product is a land community.

5   In streams, the current tends to have an equalising effect on physical factors such as pH, minerals, and oxygen.

6   The temperature of a stream may vary considerably along its course. The nearer its source, the colder it tends to be, and the less fluctuation there is. Most of the animal species colonising streams are eurythermous. Stenothermous forms tend to congregate near the source.

7   The principal effects of a current are to increase erosion of the stream bed, to maintain the water in a state of saturation with air, and to wash away most of the plankton.

8   Adaptations for survival in a current include the root systems of plants, suckers and hooks of small animals, adhesive body surfaces, streamlining, and behavioural responses.

9   Stream communities occupy a diversity of habitats including the surface film, vegetation zone, open water, and the substratum. The substratum may be

gravel, sand or mud, and is often a mixture of two or more of these. Each zone supports characteristic populations of plants and animals.

10 The successional sequence in streams (hydrosere) is similar to that in ponds but it may take place more quickly due to greater and more frequent fluctuations in the physical environment.

References

10.1 Macan, T. T. (1965) *Freshwater Ecology*, Longmans
10.2 Varley, M. E. (1967) *British Freshwater Fishes*, Fishing News Books, London

Sea shores and estuaries

For plant and animal colonists, life on the sea shore is an extension of that in the sea. The shore can be defined as that part of the coast lying between the levels of the lowest and highest tides, including the zone beyond high water which is liable only to occasional splashing with salt water. The extent and appearance of shores varies greatly according to their aspect, topography, and degree of exposure to the action of the waves. Thus, the situation in a sheltered bay can be quite different from that on an uninterrupted coastline subject to the full force of the sea and prevailing wind. For the student of ecology, the intertidal zone is one of the richest animal habitats in Britain, species of virtually every group except the higher vertebrates being represented. Plant populations are less diverse, but the absence of mosses, ferns, and flowering plants is compensated for by the abundance of algae in the form of seaweeds. While there is no such thing as an 'average' sea shore, there are certain characteristics common to most which provide excellent material for the study of ecology. It is with these features that we shall be concerned in this chapter.

The sea shore environment

The terminology used to define the different zones of a shore in relation to the rise and fall of the tide [11.1] can be rather confusing and is unnecessarily complex for ecological purposes. A somewhat simplified version is shown in Table 11.1. When applying this to an unfamiliar shore it must be remembered that non-tidal factors may also play an appreciable part in influencing the coverage and exposure of plant and animal habitats. Thus, an onshore wind coupled with low barometric pressure will result in the tide not falling as far as expected, while the opposite conditions will lead to an ebb (rise) which is higher than usual.

Within the range of tidal ebb and flow, populations of shore organisms have to withstand two cycles of tidal change every twenty-four hours. High tide is roughly fifty minutes later each day.

Gravitational effects combine fortnightly at the times of the new and full moons to cause the most extreme **spring tides**, while those occurring in between are known as **neap tides**. Tidal flow, with its ecological implications for desiccation, gas exchange, and temperature variation is the main abiotic factor influencing the spatial distribution of plants and animals on the sea shore.

From an ecological standpoint, shores can be divided into five zones (see Fig. 11.1).

(i) Splash zone
This is the region which is beyond the reach of even the highest tides. It is therefore a highly specialised land environment. The soil frequently includes blown sand and rotting remains of marine plants (seaweed) which have been washed up by the tide and blown shorewards by the wind. The terrain is

Table 11.1 Simplified scheme of nomenclature for tidal levels on a seashore

Tide level	Abbreviation	Description
Extreme high water spring tides	EHWS	The highest level reached by the sea throughout the year
Extreme low water spring tides	ELWS	The lowest level to which the tide falls throughout the year
Mean high water level	MHWL	The average level of high tide throughout the year
Mean low water level	MLWL	The average level of low tide throughout the year
Lowest high water of neap tides	LHWN	The lowest recorded level of high water during the neap tides. Below this level the shore is submerged at every tide
Highest low water of neap tides	HLWN	The highest recorded level of low water during the neap tides. Below this level the shore is not exposed at every tide of the year
Mean high tide level	MHTL	The average level of high tide throughout the year
Mean low tide level	MLTL	The average level of low tide throughout the year
Extreme low water spring tides	ELWS	The lowest level to which the tide falls during the greatest tides of the year

sometimes the forerunner of sand dunes (see Chapter 9) and is associated with rapid drying and considerable contamination by salt spray.

(ii) Upper shore
This environment is essentially marine since it is covered by the extreme high water spring tides. However, it is often out of reach of the average high tides, and the organisms inhabiting it have to withstand considerable periods of desiccation.

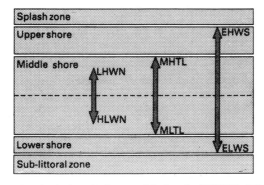

Fig. 11.1 Zonation of a shore in relation to tide levels. EHWS, ELWS = extreme high and low water springs; MHTL, MLTL = mean high and low water levels; LHWN, HLWN = lowest high-water and highest low-water neap tides (for further explanation see Table 11.1 and text)

(iii) Middle shore
This is the area covered by the average daily flow of high and low tides, being subject to exposure at both its upper and lower limits.

(iv) Lower shore
This is an extension of the lower level of the middle shore but is only exposed to the air during the extreme low water spring tides.

(v) Sub-littoral zone
This is essentially a marine environment and its substratum is never exposed even at the lowest tides. While it is not strictly part of the sea shore, it nonetheless plays an important part as a reservoir of potential plant and animal colonists.

Light

Rocky shores are usually regions of brilliant illumination as is indicated by the mass of algae that thrive there. The ecosystem is thus capable of sustaining a high level of productivity which accounts for the diversity and density of the animal populations. Sea water is frequently clear and devoid of much sediment, so even at high tide its depth is seldom sufficient to exclude light completely.

On a bright summer's day the range of light intensity in different parts of the middle shore can be striking as is shown in Table 11.2.

Table 11.2   Illumination of some localities in the middle shore zone (mid-August)

Locality	Illumination (lux)
Daylight	21 600
Rocks and rock surfaces:	
Shadow of rock (partial shade)	17 550
Shadow of rock (deep shade)	4050
Under *Fucus vesiculosus*	0
Rock pool:	
150 mm depth	14 400
300 mm depth	12 600
Under fringing *Corallina officinalis* (150 mm)	558
Under *Fucus vesiculosus* (300 mm)	0

An important point to note is the blanketing effect of large seaweeds such as *Fucus vesiculosus*. These provide protection from desiccation and overheating for many small animals such as gastropod molluscs at low tide. The humid environment beneath them also plays an important part in facilitating gas exchange by lungs or gills.

In common with many other animal populations, some of those inhabiting the sea shore, for example small crustaceans and gastropods, exhibit a distinct daily rhythm of distribution. By day they remain concealed in crevices and among the seaweeds; at night they swarm onto the surface, feeding on the abundant small algae which have been deposited by the sea. No doubt such movements have an important adaptive significance, enabling the organisms to overcome physical hazards like desiccation and also to avoid predators such as birds. Another factor besides light which causes the animals to seek shelter is wave action. Sometimes these two factors are complementary; at other times their effects are antagonistic.

In contrast to the animals inhabiting fresh water, which tend to be dull coloured, some of those on the sea shore have a brilliant colouration. An example is seen in the shells of the flat periwinkle, *Littorina littoralis*. The precise significance of this bright colouration is not known, but it must be remembered that most predators, both vertebrate and invertebrate, are colour blind, and that tone, not colour, is important in avoiding predation. In the brilliantly illuminated surface waters, which are subject to constant changes in the direction and intensity of light, high colours may play a crucial part in aiding concealment (**cryptic colouration**).

Temperature

The high specific heat of sea water makes it a vast reservoir of energy, so that such temperature changes as occur in the oceans take place only slowly. However, once the tide has receded, considerable variations exist on the shore. Some typical examples are shown in Table 11.3.

Table 11.3 Temperature of some localities in the middle shore zone (mid-August)

Locality	Temperature (°C)
Air	22.5
Sea	16.0
Rocks and rock faces:	
Shadow of rock	21.5
Under *Fucus serratus*	18.0
Rock pool:	
Surface water	21.5
200 mm depth	21.0
400 mm (bottom)	20.0

Most of the plant and animal species colonising sea shores are eurythermous p. 184) and are able to tolerate all normal changes in temperature, both diurnal and seasonal. In our offshore waters, daytime fluctuations are seldom more than a degree or so, while the maximum seasonal variation rarely exceeds 10°C. Measurements of the temperature of shore animals exposed at low tide have shown that in species such as barnacles and limpets in full sun during summer, this may reach a level 70 per cent higher than that of the sea nearby. Conditions in sand and mud are more stable, and apart from small, abrupt fluctuations in the top few centimetres, there is little change deeper down. On the shore, the greatest variations occur in rock pools and it is quite common during summer for these to be 35 per cent or more warmer than the sea (see Table 11.3). As the tide recedes, it often happens that typical inhabitants of the mud of the lower shore such as small crabs and fish like the sand goby, *Gobius minutus*, become stranded in pools higher up, to which they are not accustomed. It is significant that these species are usually the first to succumb as the temperature rises.

Humidity

With the receding tide, all organisms face the problem of desiccation; for some the situation is more serious than for others. The absence of surrounding sea poses two particular problems:

(i) a reduction in the water content of the body tissues;
(ii) an interruption in the process of gaseous exchange. Few of the colonists of the shore are air-breathers, in the majority oxygen and carbon dioxide enter and leave the body in solution in water.

Various characteristics of shore communities enable them to combat these problems:

(i) All algae possess a slimy covering of mucilage which is particularly evident in the large fucoid seaweeds which predominate in the middle shore zone. This provides an important reservoir of water. Some species such as the channelled-wrack, *Pelvetia canaliculata*, a typical inhabitant of the upper shore, have a remarkable resistance to drying. During a long period of exposure they give the appearance of brittle, black, inert masses, but once covered by the high spring tide the plants quickly assume their characteristic brown colour and rubbery consistency.

(ii) Animals that remain on rock surfaces when the tide withdraws are usually covered with a waterproof shell. Examples are the limpet, *Patella vulgata*, and the acorn barnacle, *Balanus balanoides*, which has four valve-like plates which can be closed to form a tightly fitting cover. Gill-breathing gastropods such as the edible winkle, *Littorina littorea* (Fig. 11.7 p. 222) are able, when uncovered, to close the branchial chamber with a horny plate called the operculum. (They are known as **operculates** in contrast to lung-breathing species which are **pulmonates**.)

(iii) Many vulnerable species overcome the problem of desiccation by appropriate behaviour, and move into the shelter of rock crevices or underneath the carpet of seaweeds. Populations of soft-skinned animals such as the sea anemone, *Actinia equina*, thus exhibit different patterns of spatial distribution depending on the state of the tide. One of the most remarkable of these behaviour patterns is the homing ability of limpets (*Patella*). At night they roam over the surface of rocks, rasping with their radulas at the small algae which encrust the surface. By day, they return to their former site which may be a metre or more away. So precise is their pattern of movement that on soft rocks 'home' is often marked by a ring-shaped depression (limpet scar) which conforms to the outline of the shell when pressed downwards (Fig. 11.2). Our understanding of the homing mechanism is still incomplete but the behavioural pattern presumably ensures that the animal maintains the position on the shore to which it is best adjusted. The exact fit between shell and rock serves to reduce water loss and also the likelihood of being prised off by predators such as gulls.

(iv) Special problems face the occupants of sand and mud that are frequently found in the lower shore zone. Many of these such as the lugworm, *Arenicola marina*, a polychaete worm, are soft bodied and are unable to withstand exposure at the surface. Bivalve molluscs such as the cockle, *Cardium edule*, gain protection by closing their shells but are unable to remain in this condition for long.

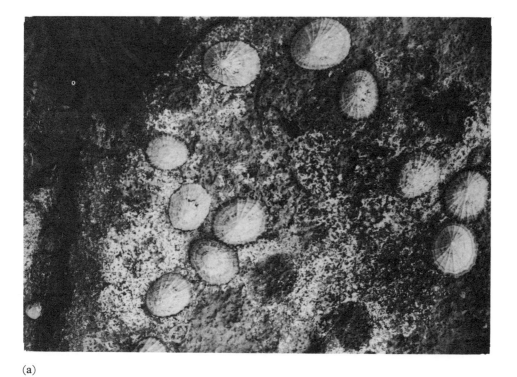

(a)

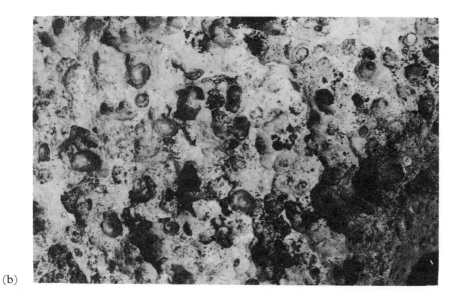

(b)

Fig. 11.2 (a) Limpets, *Patella vulgata*, on a rock surface with two recent scars, (b) limpet holes and scars in soft lias rock

(v) Another approach to water conservation is the physiological one, notably the avoidance of undue loss due to excretion. The major nitrogenous end-product of the metabolism of most aquatic animals is ammonia, which is highly soluble and toxic. It has to be removed as quickly as possible and this requires large quantities of water. In land vertebrates with a problem of water shortage such as birds, snakes, and lizards, ammonia is converted into non-toxic and relatively insoluble uric acid. This is also true of some sea-shore invertebrates such as gastropods of the genus *Littorina*, in which those with a greater capacity for producing uric acid have a greater ability to colonise the drier parts of the shore. Of the four common species (Fig. 11.7 p. 222), *L. neritoides* is almost confined to the upper shore, *L. littorea* extends throughout the middle and lower shore and the other two species, *L. saxatilis* and *L. obtusata* occur in between. Needham examined the uric acid content in the excretory organs of the four species of *Littorina* [11.2] and showed that this conforms closely to their distribution on the sea shore (see Table 11.4).

Table 11.4   Uric acid content of the excretory organs of four species of *Littorina* in relation to their position on the shore, estimated by Needham (after Nicol)

Species	Uric acid (mg g^{-1} dry mass)	Position on shore
Littorina neritoides	25.0	High
L. saxatilis	5.0	↓
L. obtusata	2.5	
L. littorea	1.5	Low

Zonation on the sea shore

(a) Plants

The physical conditions likely to be encountered by plants and animals attempting to colonise a rocky shore are some of the most extreme that exist in any range of habitats. The most important of these have been considered in the previous section. Different species are able to adapt to such conditions in varying degrees according to their ecological requirements and this results in a pattern of distribution characterised by a distinct **zonation** related to the rise and fall of the tide. In attempting to interpret the situation it is necessary to think both horizontally and vertically, hence the need to survey the part of a shore to be investigated (see p. 120). On gently sloping shores (Fig. 11.3) the zonation of organisms between low and high tide marks will be reasonably clear. But on more discontinuous shorelines with large boulders, an equally clear-cut pattern of distribution can often be observed on the vertical faces of the large rocks. It is not easy to generalise regarding the sequence of species as this can vary considerably from one shore to another. A fairly typical pattern is illustrated in Fig. 11.4. Predominant among the plants at the seaward edge are the various oar-weeds such as *Laminaria digitata* (Fig. 11.5(a)), together with the serrated wrack, *Fucus serratus* (Fig. 11.5(b)), which occurs on all except the most exposed coasts. Higher up the shore is the knotted wrack, *Ascophyllum nodosum* (Fig. 11.5(c)) which is

Fig. 11.3 A typical gently-sloping rocky shore at about half-tide level (photograph D. P. Wilson/Eric and David Hosking)

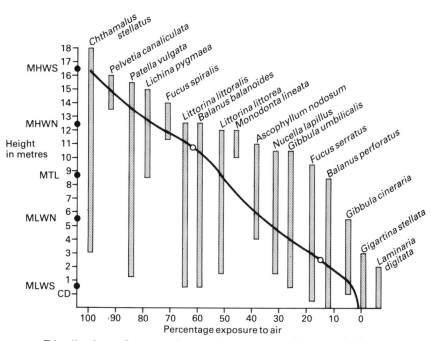

Fig. 11.4 Distribution of some plant and animal species in relation to tide level on a south Devon shore. The line indicates the percentage of time that the shore is exposed to the air at each level (From Southward, A. J. (1965) *Life on the Sea Shore*, Heinemann)

(a)

(b)

(c)

Fig. 11.5 Some common seaweeds of the middle and lower shore. (a) Oar-weed, *Laminaria digitata*, exposed at extreme low tide. (b) Serrated wrack, *Fucus serratus*. (c) From top to bottom, channelled wrack, *Pelvetia canaliculata*; flat wrack, *Fucus spiralis*; knotted wrack, *Ascophyllum nodosum* (photograph D. P. Wilson/Eric and David Hosking)

widespread and abundant and whose distribution frequently overlaps that of *F. serratus* below, and the flat wrack, *F. spiralis* (Fig. 11.5(c)) above. Still further up the shore and extending to the level of the highest tides is the channelled wrack, *Pelvetia canaliculata* (Fig. 11.5(c)). Beyond this again we come to the splash zone (see Fig. 11.1) which is seldom, if ever, reached by the tide but is subject to a good deal of salt spray. Here we find a peculiar land flora some of which are salt-tolerant (halophytes) and others drought-tolerant (xerophytes) including numerous lichens (e.g. *Rhizocarpon* sp) and flowering plants with succulent leaves such as samphire, *Crithmum maritimum*, and golden samphire, *Inula crithmoides*. (Many animals inhabit rotting seaweed, most being insects breathing atmospheric air.) The sequence of zonation is summarised in Fig. 11.6.

(b) Animals
Like the plants, the animals inhabiting the rocky shore are well adapted to withstand exposure, and many possess effective devices for attachment to rocks or the fronds of seaweeds, for example the muscular foot of gastropod molluscs.

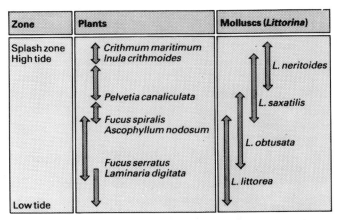

Fig. 11.6 Zonation of some common plants and animals (molluscs) on a rocky shore

These animals also exhibit a characteristic zonation, typified by the distribution of the various species of periwinkle, *Littorina* sp (Figs 11.6 and 11.7). Thus the lower part of the shore is dominated by the common periwinkle, *L. littorea*, which extends over the zones of *Fucus serratus*, *Ascophyllum* and *F. spiralis*. Slightly higher up the shore and overlapping the distribution of *L. littorea* to some extent is the flat periwinkle, *L. obtusata*. The *Pelvetia* zone is colonised by two further species, the rough periwinkle, *L. saxatilis*, which often extends downwards into the region of *Fucus spiralis*, and the small periwinkle, *L. neritoides*. The latter could be regarded with some justification as a partial land animal since it can exist for days in the splash zone well above high-tide mark. Similar zoned distributions occur in many other shore animals, particularly among different species of top shells (Gastropoda) of the genera *Gibbula* and *Calliostoma*.

Investigating zonation
In the previous chapters it has been emphasised that when studying a complex community, there is much to be said for selecting a few species for detailed investigation rather than attempting a wider and more superficial survey. This is particularly true of the rocky shore and highlights the value that can be gained from the study of small bays and inlets. Some of the results of such a study are summarised in Fig. 11.8. These results are derived from a belt transect 14 metres long and a metre wide. The plants were scored for each square metre as rare (less than 25 per cent of the stones covered), common (25–75 per cent coverage), and abundant (more than 75 per cent coverage). The animals were counted as individuals per square metre. Among the plants there was a clear-cut zonation, the presence of *Enteromorpha intestinalis* and *Ulva lactuca* indicating that an appreciable amount of fresh water must have been draining onto the shore from the land behind, since both are brackish species. The molluscs too were clearly zoned, the distribution of *Littorina neritoides* being of particular interest. Far from being confined to the upper reaches of the shore (see Fig. 11.6) it had colonised almost the whole of it, perhaps as a result of the lack of competition by other species.

Before leaving zonation it is worth considering whether any direct experimental evidence exists in support of the view that particular abiotic factors such as desiccation, either acting separately or together, play a predominant role in influencing the pattern of plant and animal distribution on the sea shore. Experimental data are not easy to find and such as exist appear to have been obtained many years ago. Southward [11.1] quotes the results of experiments carried out on young plants of fucoid seaweeds. Thus *Fucus spiralis*, which occurs towards the upper part of the shore (Fig. 11.6 p. 220 and Fig. 11.5 p. 219) was found to grow best under conditions where six hours of wetness were followed by six hours of dry. Little growth occurred when it was either kept out of water or totally immersed for eleven out of twelve hours. By contrast, *F. serratus* which occurs low down on the shore, grew best when almost continuously immersed but did not flourish when six hours immersion alternated with six hours of drying, and died when kept out of water for eleven out of twelve hours. *Ascophyllum*, a mid-tide species, gave intermediate results, showing some growth when submerged for eleven out of twelve hours and under equal periods of immersion and drying. But it failed to grow altogether or died when desiccated for eleven out of twelve hours. In general, therefore, spatial distribution and the physiological requirements of these plants seem to be correlated.

Among animals, as we have already seen, there is indirect evidence of a comparable correlation in the form of a relationship between the production of uric acid as an excretory product in periwinkles (*Littorina*) and the capacity of different species to withstand varying degrees of exposure (p. 214). As long ago as 1910, Colgan [11.3] measured the maximum survival time of littoral molluscs when removed from their normal habitat and exposed to dry air. His results are summarised in Table 11.5.

Table 11.5 Survival of four species of *Littorina* (Gastropoda) in dry air (after Colgan)

Species	Period of survival in dry air (days)
Littorina neritoides	42
L. saxatilis	31
L. obtusata	6
L. littorea	23

The periods of survival of *Littorina neritoides* and *L. saxatilis* are comparable with their respective periods of exposure relatively high up on the sea shore. The findings for *L. littorea* present something of an anomaly in that the species appears to be roughly four times as tolerant of desiccation as *L. obtusata* although it occupies a more seaward zone on the shore. A possible explanation may lie in the close association of *L. obtusata* with the large seaweeds such as *Fucus spiralis* where there is a plentiful supply of moisture and an abundance of oxygen due to photosynthesis. *L. littorea* frequently inhabits sandy and muddy localities which tend to be devoid of the shelter provided by extensive vegetation, hence the advantage of its greater ability to withstand exposure. Finally, it should be borne in mind that the figures for survival summarised in Table 11.5 represent the maximum duration of life, and it is probable that serious harm would have resulted from much shorter periods away from water.

(a)

(b)

(c)

(d)

Fig. 11.7 Four species of periwinkle (Gastropoda) with a distinct zonation on the sea shore (a) common periwinkle, *Littorina littorea*; (b) flat periwinkle, *L. obtusata*; (c) rough periwinkle, *L. saxatilis*; (d) small periwinkle, *L. neritoides*, just above high water (photograph D. P. Wilson/Eric and David Hosking)

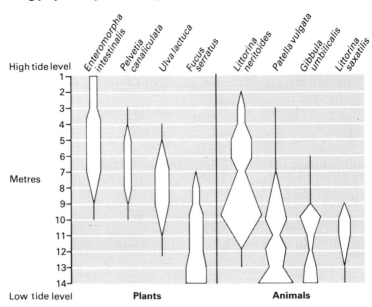

Fig. 11.8   Distribution of eight species (four plants and four animals) on a small shore

Habitats on the sea shore

The intertidal zone is an area rich in plant and animal life. Most of the species that live there are permanent colonists but these are regularly reinforced by the tide. As it retreats it leaves behind fragments of algae and numerous animals, many of which become trapped in rock pools and elsewhere. It also deposits a massive quantity of plankton and detritus which forms a film on the rock surfaces and the seaweeds covering them, providing food for a wide range of small animals. Some typical habitats are summarised in Fig. 11.9. Each habitat demands particular adaptations if it is to be successfully colonised. The lower shore frequently consists of sandy beaches, which can provide a useful ecological study on their own provided ample time is available. Most of the animals such as the lugworm, *Arenicola marina* (Fig. 11.16 p. 231 and Fig. 11.10), which live here are burrowers and this enables them to avoid desiccation at low tide. The majority feed on detritus or plankton. Some possess beautiful adaptations for gas exchange, for example, the bivalve molluscs *Cardium edule* and *Tellina tenuis* (Fig. 11.11), in which the length of the siphon determines the depth at which the animal can live.

The food relationships on the sea shore are complex and some typical ecological niches and their relationships are shown in Fig. 11.9. Comparatively few herbivorous animals depend upon the larger plants for their food, the majority feeding on small fragments and minute algae deposited from the plankton and encrusting the surfaces of rocks and seaweeds. Many, including some gastropod molluscs, all bivalves, and the smaller crustaceans, annelids, coelenterates and sponges are filter feeders. Among the primary consumers, crustaceans in particular are an important source of food for the smaller carnivores, while molluscs are a major item of diet for fishes and birds.

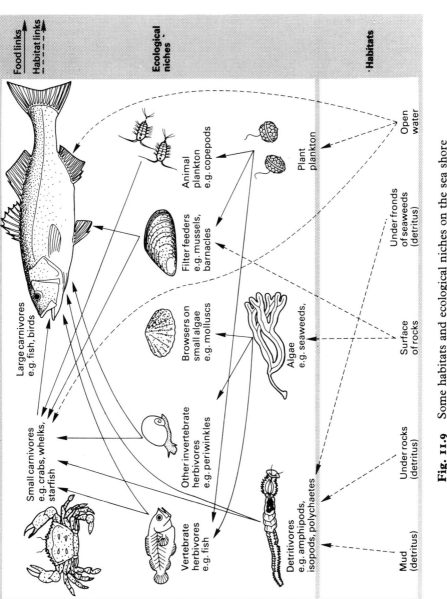

Large carnivores
e.g. fish, birds

Animal
plankton
e.g. copepods

Plant
plankton

Open
water

Filter feeders
e.g. mussels,
barnacles

Under fronds
of seaweeds
(detritus)

Browsers on
small algae
e.g. molluscs

Algae
e.g. seaweeds,

Surface
of rocks

Small carnivores
e.g. crabs, whelks,
starfish

Other invertebrate
herbivores
e.g. periwinkles

Under rocks
(detritus)

Vertebrate
herbivores
e.g. fish

Detritivores
e.g. amphipods,
isopods, polychaetes

Mud
(detritus)

Fig. 11.9   Some habitats and ecological niches on the sea shore

Fig. 11.10 A muddy shore heavily colonised by the lugworm, *Arenicola marina*. Note the casts and headshaft holes (see also Fig. 11.16 p. 231) (photograph D. P. Wilson/Eric and David Hosking)

Some topics for investigation

1 Set up a belt transect (p. 89), one metre wide, from the splash zone to the lower shore, (or covering part of this distance). In order to gain a valid idea of the relationship with the tide it is desirable to survey the shore (or part of it) first (p. 120). Select a few species (say five plants and five animals) for quantitative sampling if the terrain permits. The plants can be estimated as percentage cover of the substratum, the animals as number per square metre. Construct a distribution chart (Fig. 11.8). What evidence is there of a zonation of species? Apart from exposure to the air at low tide, what other factors could

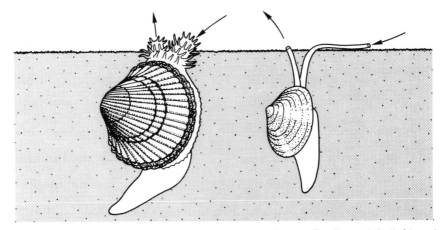

Fig. 11.11 Two common sand-burrowing bivalve molluscs, *Cardium edule* (left) and *Tellina tenuis* (right). The length of the siphon is adapted to the depth at which the animal lives (after Yonge)

be influencing the pattern of spatial distribution (for example, a flow of fresh water from the land)?

2 Compare two beaches, one exposed and the other sheltered. To what extent can plant and animal distribution be related to the different conditions?

3 In a portion of a sea shore community, what adaptations exist for withstanding the eroding effects of wave action, (a) among plants, (b) among animals?

4 In a rock pool, what evidence is there of a zonation comparable to that on the rest of the shore? To what extent can the animals present be regarded as (a) permanent colonists, (b) temporary visitors? (If time permits, any hypotheses could be checked by marking experiments. Remember that a surface must be dry before the application of paint marks.)

5 In a rock pool exhibiting plant (and animal) zonation, can the sequence of stages be correlated with (a) temperature changes, (b) variation in illumination?

6 Limpets (*Patella* sp) browse, when covered by the sea, on the small algae covering the rocks in their immediate vicinity, but as the tide retreats they return to their 'homes' (identifiable as ring-like depressions on the rock surface). To what extent does this browsing action of limpets influence the distribution of plant species in their immediate vicinity?

7 In the splash zone, an accumulation of drying seaweeds often provides a habitat for a large community of insects and small crustaceans. Make a collection of these and identify them as far as possible (at least down to class). How do (a) their mode of behaviour, (b) their methods of gaseous exchange differ from animals occupying the adjacent shore? Do the flowering plants of the splash zone show any evidence of adjustment to halophytic (salty) conditions, for example, succulence due to increased water storage tissue?

8 Marine plankton, like that in fresh water, tends to be most active at night. Obtain samples after dark from (a) a rock pool, (b) the sub-littoral zone of the shore, using a standardised sampling procedure with a plankton net (see p. 73). Make a rough comparison of the densities in the two localities using a cavity chamber (p. 93). How do the densities compare? Can you account for any differences in (a) the kinds of organisms present, (b) their numbers?

9 As a long-term study, scrape clean a number of rocks at different levels on the sea shore, marking each with paint (p. 90) when the surface is dry so that it can be identified. Follow the subsequent process of recolonisation. What is the sequence of succession? How is the succession of plants related to that of animals? Does the sequence vary at different levels on the shore?

The estuarine environment

An estuary is essentially a river valley which has been invaded by the sea. From an ecological standpoint, therefore, it represents an extension inland of a sandy shore. Its two main physical features are:

(i) considerable expanses of soft mud which are often exposed at low tide;
(ii) a brackish zone of widely fluctuating salinity where the flow of the tide inland meets that of the river seawards. High tide is a period of maximum salinity while at low tide and during periods of heavy rain the range of the fresh water zone is greatly increased.

The problems facing potential plant and animal colonists are thus of two main kinds: adaptation to the peculiar physical environment of mud, and adjustment to osmotic changes resulting from fluctuations in the salt content of the water. Such problems are not found in any other habitat in Britain and this makes the study of estuarine ecology particularly worthwhile.

Life in estuaries

The plant populations bordering estuaries comprise a peculiar salt-tolerant community. Such species are known as **halophytes** and exhibit a range of adaptations for maintaining their internal osmotic balance which are comparable with those occurring in colonists of the splash zone at the highest level of the sea shore (p. 210). The most obvious of these is the development of succulent (water storage) tissue, as is found in glasswort, *Salicornia* sp, a frequent inhabitant of the mud in the vicinity of the high tide mark (Fig. 11.12). Similar tendencies towards increased succulence are found in such species as the sea aster, *Aster tripolium*, and the sea purslane, *Halimione portulacoides*. Many other halophytic species have overcome the problems of osmotic balance by maintaining the concentration of solutes in the cell sap at a relatively high level, thereby enabling some water uptake to occur from the surrounding saline medium.

Although mud provides a good physical medium for plant roots, the fine compacted particles lead to anaerobic conditions, which become evident only a few centimetres below the surface. Many species such as the cord grass, *Spartina townsendii* (Fig. 11.13), possess aeroid tissue which provides a reserve of air and facilitates gaseous exchange during periods of exposure.

Relatively few animal species have been able to adjust to the peculiar ecological conditions of estuaries. However, when they occur, the populations frequently attain a high density. This is due not so much to the productivity of the estuarine zone as to the abundance of organic material that is washed inwards from the sea at

Fig. 11.12 Glasswort, *Salicornia* sp, a succulent plant species colonising the high-tide zone of estuaries

Fig. 11.13 Cord grass, *Spartina townsendii*, a colonist of estuarine mud flats. Its rhizomes contain aeroid tissue and serve the triple functions of gas exchange, anchorage, and vegetative growth

high tide and outwards from the river at low water. Some fishes such as the salmon are essentially marine but move up river to spawn, while eels mature in rivers but breed in the sea. Although these fish have achieved considerable powers of osmoregulation, they must, nonetheless, be regarded as only temporary inhabitants of estuaries. Most marine animals are incapable of penetrating far into the brackish zone since their capacity for osmoregulation is strictly limited (**stenohaline** species). Those that have succeeded have all achieved some degree of osmotic independence (**euryhaline** species). This has been done in two ways:

(i) *Adjustment of osmotic control*

For marine species adjusting the mechanism of osmotic control has necessitated the evolution of a means of controlling the passage of water into their bodies which would otherwise occur. Similarly, for freshwater species colonising estuaries, the problem is to overcome a loss of water.

Among fishes, the majority are stenohaline and unable to survive in estuaries. An exception is the flounder, *Pleuronectes flesus*, which is capable of swimming some way up river and even of surviving in fresh water. Evidently, its scales provide an almost impermeable covering and any excess water entering the body can be disposed of through the kidneys. The distinction between the adaptedness of stenohaline and euryhaline species is well illustrated from the data of Baldwin [11.4] on the osmoregulating capacity of the spider crab, *Maia*, and the shore crab, *Carcinus*. Reference to Fig. 11.14 shows that whereas in *Maia* (a marine species), the relationship between changes in external and internal osmotic pressure is linear, in *Carcinus* (an estuarine species), a considerable change in

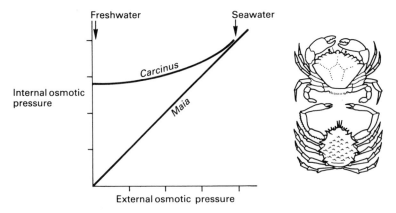

Fig. 11.14 Variation in the internal osmotic pressure with changes in external osmotic pressure in two species of crabs (after Baldwin)

salinity is needed before an appreciable alteration in internal osmotic pressure occurs.

In general, crustaceans appear to be particularly able to adapt to conditions in estuaries and the salt marshes that are frequently associated with them. Thus the chameleon shrimp, *Praunus* sp, is often abundant in open water, while the prawn, *Palaemonetes varians*, and the isopod, *Sphaeroma rugicauda*, form large populations in pools and gulleys that are isolated at low tide. One of the best indicators of changing salinity is the amphipod, *Gammarus*, three of whose species cover a range from the open sea to fresh water (Fig. 11.15). Thus *G. locusta* is a stenohaline species incapable of survival away from the sea and *G. pulex* is also stenohaline and is confined to fresh water. Euryhaline species include *G. zaddachi* whose distribution overlaps that of the other two, extending to the sea and also far up river.

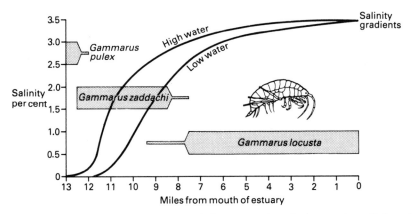

Fig. 11.15 Distribution of three species of *Gammarus* in relation to the salinity of an estuary (after Yonge)

(ii) Burrowing in mud

Animals can avoid the most extreme variations in salinity by burrowing. The majority of burrowing species possess no great powers of osmoregulation but, nonetheless, some are successful colonists of estuaries. Typical among the annelids is the lugworm, *Arenicola marina* (Fig. 11.16), whose burrows (Fig. 11.10 p. 226) enable it to descend to a sufficient depth for the effects of salinity variations to be minimal. Its range often extends up river for a considerable distance, reaching areas subject to the maximum changes in salinity occurring between high and low water. Another common burrowing polychaete is the ragworm, *Nereis diversicolor*, which also inhabits the muddy reaches of many estuaries. Among bivalve molluscs, the oyster, *Ostrea edulis*, frequently penetrates up river, as does the cockle, *Cardium edule*, whose respiratory siphon is well adapted for a burrowing existence (Fig. 11.11 p. 226).

Succession in estuaries

The continual movements of silt and mud occurring in estuaries are comparable to those in rivers described in the previous chapter, but whereas in inland waters they take place in one direction only, in estuaries the movement is two-way and takes place with a regular periodicity. Where obstructions occur, mud banks quickly accumulate, and if these are in mid-stream they result in the formation of islands. More usually, the gradual accumulation of mud occurs round the stems of plants such as the cord grass, *Spartina townsendii* (Fig. 11.13), which frequently grows in dense masses along the margins of estuarine mud flats. As a result, the substratum gradually rises above water level and this is accompanied by corresponding changes in the plant and animal communities.

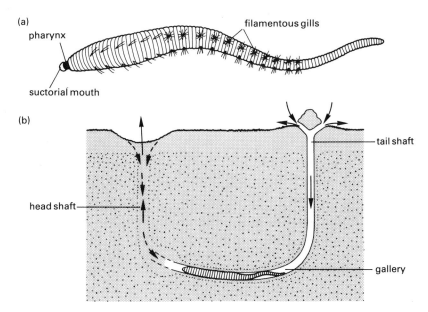

Fig. 11.16 (a) the lugworm, *Arenicola marina*, (b) the structure of its burrow which may penetrate 20 cm or more below the surface, affording considerable protection from osmotic changes in the water above (after Yonge). (See also Fig. 11.10 p. 226)

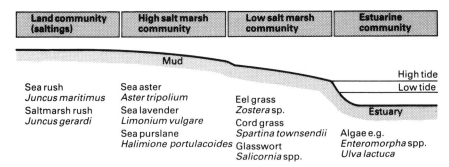

Fig. 11.17 Plant succession in an estuary (halosere)

The process of succession in estuaries resembles that of the hydrosere described earlier (p. 196) but here any potential plant colonist has to face the two additional problems already mentioned, namely, the instability of the mud and the high concentration of salt. As we have seen, many species capable of surviving in such conditions exhibit characteristic features such as the possession of succulent tissue for water retention, and are known as halophytes. The succession, water – salt marsh – land, is sometimes referred to as a **halosere**. The sequence of steps by which these changes come about and the rate at which they occur vary considerably from one locality to another, so the diagram (Fig. 11.17) provides no more than a generalised view of the process.

The larger aquatic plants inhabiting estuaries are exclusively algae. Near the mouth where the salinity is high, they may include various fucoid seaweeds, but further up river these are replaced by such species as the sea lettuce, *Ulva lactuca*, and the various species of *Enteromorpha*. Bordering the water is a community of plants (the low salt marsh community) whose roots are usually submerged in water at high tide but which may have to withstand days of continuous exposure during the low spring tides. Species typical of this zone include those which are so attractive to migrating geese in the vicinity of the Severn estuary, the cord grass, *Spartina townsendii* (Fig. 11.13), the glasswort, *Salicornia* sp. (Fig. 11.12), and the eel grass, *Zostera* sp. Further inland the high salt marsh community includes characteristic species such as the sea aster, *Aster tripolium*, the sea lavender, *Limonium vulgare*, and sea purslane, *Halimone portulacoides*. Inland again, succession leads to a land community, albeit with a tolerance of salt (hence the term 'saltings'). The moister regions support a number of rush species such as the sea rush, *Juncus maritimus*, and salt marsh rush, *J. gerardi*, while the whole area is colonised by a variety of grasses which provide good grazing for cattle.

Unlike many rocky shores, the vertical distance covered by a halosere is usually quite small. Thus in an estuarine creek studied by Humphreys [11.5] and illustrated in Fig. 11.18, this amounted to no more than 4 metres in a horizontal distance of 42 metres.

Few estuarine and salt marsh animals rely directly on the plants growing in these places for their food. As was pointed out earlier, mud contains an abundance of organic debris either washed seawards by the river, or inland by the sea, or derived from the decomposition of the dead remains of local plants. Detritus thus provides the staple diet for most of the animal populations, the majority of which

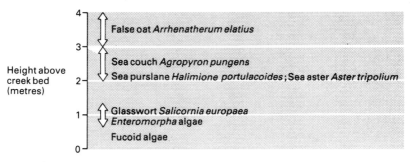

Fig. 11.18 Succession of some estuarine plant species in relation to height (data from Humphreys)

utilise filter feeding mechanisms in some form or other. Burrowing species such as the cockle, *Cardium edule* (Fig. 11.11 p. 226), the Baltic tellin, *Macoma baltica*, the lugworm, *Arenicola marina* (Fig. 11.16 p. 231), and the catworm, *Nephthys hombergi*, often form dense populations in the intertidal mud flats, and when the tide recedes these provide food for large flocks of birds (waders). Some annelids such as the ragworm, *Nereis diversicolor*, are partially carnivorous, feeding on small crustaceans and larvae. Inland, pools and creeks frequently support great numbers of crustaceans such as the prawn, *Palaemonetes varians*, and the isopod, *Sphaeroma rugicauda*, both typical filter feeding species. The environment facing these animals is an exacting one. Not only are they required to withstand the normal range of fluctuations in estuarine salinity, but at low tide and during warm weather the salt concentration in pools may increase considerably due to evaporation. Again, shallow areas of water can be subject to great variations in temperature, while prolonged exposure during the lowest spring tides may result in their drying up altogether.

Postscript: estuaries and motorways

At first sight, the relationship between the conditions in estuaries and those on motorways would seem to be remote. However, the work of Scott and Davison [11.6] has revealed that since the 1960s no less than ten typically halophytic plant species have found their way into the corridor of salt-contaminated ground beside a number of our major roads. It is estimated that around one and a half million tonnes of salt are now applied to the roads of Britain annually and in places, this has produced environments comparable in many ways to those discussed earlier in this chapter. One possible point of access for seeds of estuarine plants seems to have been in Northumberland where the road network runs close to salt marshes. Thence, seeds could have been carried on car bodies and tyres down the A1 and then to the M1 motorway as far south as Bedfordshire. A similar spread of halophytes has been recorded from Kent along the A2, M2 and M20 as far as the Isle of Sheppey. Colonists include the sea plantain, *Plantago maritima*, sea spurrey, *Spergularia rubra*, and sea aster, *Aster tripolium*, all typical salt marsh species.

The remarkable distribution of these plants through the agency of motor cars parallels the equally extraordinary spread late last century of the Oxford ragwort,

Senecio squalidus, in the wake of passing trains. In this way its seeds were transported from the city of its name to railway cuttings throughout the country.

Finally, it is noteworthy that this motorway phenomenon is not peculiar to Britain. In West Germany the map of the inland distribution of salt marsh plants such as the sea meadow grass, *Puccinellia maritima*, closely duplicates the pattern of the autobahn system. A similar spread of halophytic species along main roadways has been recorded in France, Holland, Switzerland, and East Germany.

Some topics for investigation

1   The sequence of plant succession in estuaries varies from one locality to another. Select two (or more) different localities for comparison using the belt transect method. How do you account for any variations observed? (Note: It is important to take account of the vertical distance as well as the horizontal, so each site should be surveyed, if possible.)

2   In **1** above, is there any relationship between the number of plant species exhibiting succulent features in their stems and leaves, and their distance inland from the high water mark of the estuary?

3   The density and variety of animal populations in estuarine mud pools varies greatly. Sample a number of different pools and compare them in terms of (a) numbers of species, (b) population density, (c) salinity (p. 118). To avoid a time-consuming estimation of numbers, also to facilitate comparison, it may be convenient to devise arbitrary scales of abundance such as those suggested by Tait [11.7]:
A – abundant   C – common   F – frequent   O – occasional   R – rare
N – not found
A separate set of numerical values for each category will probably be needed for different species.

4   Is it possible to correlate the results obtained in **3** above with variations in (a) salinity, (b) temperature, (c) oxygen content of the water?

5   The roots of plants such as cord grass, *Spartina townsendii*, are said to withstand anaerobic conditions in the estuarine mud. To what extent is this so? (Also in other species such as glasswort, *Salicornia* spp.) Do sections through the rhizomes of *Spartina* show evidence of aeroid tissue?

6   Using one metre quadrats (p. 86) (or any other appropriate size) sample a population of burrowing animals such as the lugworm, *Arenicola marina*, in the intertidal zone, either by counting the number of casts or by digging. Do variations in density conform to any particular pattern? If so, can you relate variations to particular ecological factors such as (a) size of mud particles, (b) organic content of mud, (c) period of exposure by the tide?

SAFETY

Mud can be treacherously soft so you need to be careful how you walk. The shallowness of estuaries greatly increases the speed of the incoming tide, and if you are not alert it is easy to get cut off!

Summary

1 The relationship between the zones of the shore and changes in the tide is described by a standard terminology.

2 From an ecological standpoint, shores can be divided into five zones. From the land seawards these are the splash zone, upper shore, middle shore, lower shore, and sub-littoral zone.

3 Light intensity can vary greatly on a shore and plays an important part in influencing animal distribution.

4 Temperature variations are not as great or as rapid as those of light. Their effects are greatest on animals living in rock pools when exposed by the tide.

5 The capacity of sea-shore plants and animals to withstand exposure at low tide and a consequent reduction in humidity, varies greatly. Evidence of this is provided by the characteristic zonation of populations.

6 Extraneous factors such as streams of fresh water flowing seawards from the land can exert a profound effect on the pattern of zonation in shore communities.

7 The sea shore is rich in plant and animal life, and the intertidal zone provides good opportunities for the study of a diversity of habitats ranging from rocks to mud. The ecosystems involved are usually complex.

8 The two main features of estuaries are expanses of soft mud, which are exposed at low tide, and brackish water of varying salinity.

9 Plants colonising the estuarine zone (halophytes) exhibit a variety of adaptations for a saline environment. Animals have to withstand considerable variations in salinity and the problems of osmoregulation that these entail. Most have a wide osmotic tolerance (euryhaline).

10 Animal species with a narrow osmotic tolerance (stenohaline species) occur either at the mouth of an estuary (near the sea) or up stream in fresh water. Some avoid variations in salinity by burrowing in mud.

11 Succession in estuaries resembles that of a freshwater hydrosere. The steps in the sequence and the rate at which they occur vary from one locality to another.

12 Estuarine animals tend to rely for their food on organic debris and plankton washed up by the tide. Few eat estuarine plants.

13 The salting of motorways in winter has produced halophytic conditions alongside them. These have now been colonised by a variety of estuarine plants whose seeds must have been transported inland by vehicles travelling from the coast.

References

11.1 Southward, A. J. (1965) *Life on the Sea Shore*, Heinemann

11.2 Nicol, J. A. C. (1961) *The Biology of Marine Animals*, Pitman

11.3 Colgan, N. (1910) 'Factors affecting survival of littoral molluscs', *The Irish Naturalist*

11.4 Baldwin, E. (3rd edn 1949) *An Introduction to Comparative Biochemistry*, Cambridge

11.5 Humphreys, T. J. (1981) 'Estuarine ecology as project work', *Journal of Biological Education*, **15**, 225–33

11.6 Scott, N. E. and Davison, A. W. (1982) 'De-icing salt and the invasion of road verges by maritime plants', *Watsonia*, **14**, 41–52

11.7 Tait, R. V. (3rd edn 1981) *Elements of Marine Ecology*, Butterworths

Heaths and bogs

Large areas to the north and west of Britain, also more restricted localities in the south, are dominated by a type of terrain known as **heathland**. Its chief characteristic is uniformity (Fig. 12.1), the vegetation being predominantly of two species, ling, *Calluna vulgaris*, and bell heather, *Erica cinerea* (Fig. 12.2). In windswept and exposed areas where trees are unble to establish themselves, heath represents a climax vegetation. Elsewhere, it may owe its formation and continuing existence, either directly or indirectly, to the influence of man, through the effects of burning and grazing. In the low ground where drainage is poor, standing water tends to accumulate and this results in the formation of peat **bogs**. These support a characteristic flora which includes insectivorous species such as the sundew, *Drosera* sp. (Fig. 12.6 p. 242) and other indicators like cotton grass, *Eriophorum angustifolium* (Fig. 12.9 p. 245) and bog moss, *Sphagnum* sp.

From an ecological standpoint, the uniqueness of heaths and bogs is their greatest attraction, since survival in such peculiar conditions necessitates adaptations by plants and animals some of which are not found elsewhere. For practical purposes, the relatively few species represented is also an advantage since this greatly aids the problem of identification.

Fig. 12.1 Heathland uniformity (Dunkery Beacon, Exmoor). The predominant plant is ling, *Calluna vulgaris*. The area is windswept and trees are absent. The patch of heath rushes, *Juncus squarrosus*, in the foreground indicates damp conditions (wet heath)

Fig. 12.2   Bell heather, *Erica cinerea*, and ling, *Calluna vulgaris*, growing on a dry heathland slope

The formation of heaths and bogs

In order to understand the ecology of heath and bog ecosystems, it is necessary to have some idea of the ways in which they have come about. Reference to these has already been made in Chapter 8 (p. 151), where it was pointed out that soils can be divided roughly into two types, **mull** which includes the alkaline and neutral conditions found in much of Britain and **mor** consisting of the acid soils that comprise the heathlands with which we are concerned here. Two main factors combine to produce a mor soil: a porous medium such as sand or gravel and a relatively high rainfall. Valuable solutes present in the surface layers tend to be washed downwards (leached) into the parent rocks below. The aerobic bacteria which break down dead plant remains and convert them to humus (causing decay), flourish in an approximately neutral or alkaline medium. But leaching tends to wash the bases out of the surface zone producing acid conditions with a pH of around 4.0. This has the effect of slowing down bacterial action, so that an organic layer consisting of incompletely decomposed plant remains (**peat**) results. Leaching also has the effect of washing minute particles of iron and organic compounds downwards and these accumulate to form a hard impervious layer called a **pan**. A soil zoned in this way is known as a **podsol** (Fig. 12.3 and Fig. 8.1 p. 150) and is characteristic of heaths. Its main features are rapid drainage and therefore low humidity, high acidity, and a lack of nutrient minerals.

In low-lying areas, waterlogging tends to occur. This is due partly to the impervious nature of pans and also to the accumulation of clay, which is

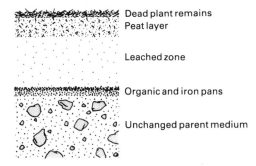

Dead plant remains
Peat layer

Leached zone

Organic and iron pans

Unchanged parent medium

Fig. 12.3   Section of a heathland soil (podsol). (See also Fig. 8.1 p. 150)

frequently an additional feature. Such conditions become increasingly anaerobic, thus inhibiting bacterial decay. Anaerobic bacteria take over, forming a foul-smelling, acid, organic mud, which may be several feet deep, overlain by a layer of waterlogged peat. Pearsall [12.1] obtained the following values for pH at different levels of **moorland** (a blanket term covering heaths and bogs) in the Pennines.

Table 12.1   Variation in the acidity (pH) of moorland soils in relation to plant types (after Pearsall)

Locality	Level	Indicator plant species	pH
Heath	Driest	Bilberry, *Vaccinium myrtillus*	2.8–3.3
Heath	Dry	Heather, *Erica cinerea*	3.4–3.7
Heath	Moist	Moor sedge, *Carex binervis*	3.2–3.8
Bog	Mud	Bogmoss, *Sphagnum* sp.	4.7–5.4

In this instance the pH of the soil increased with decreasing height and increasing humidity. One explanation could be that the bases which had been leached out of the surface layers by the rain had tended to accumulate lower down.

The sequence of stages in the formation of heaths and bogs outlined above is summarised in Fig. 12.4.

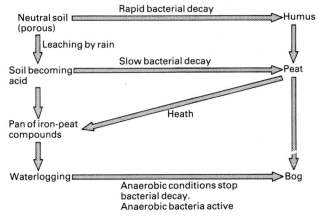

Fig. 12.4   Stages in the formation of heaths and bogs

The heath community

Plants

All heathland species are tolerant of acidity and the majority are intolerant of lime (**calcifuge**). Due to the exposed situation in which most heathland occurs, much of it is devoid of trees and this helps to reinforce its general appearance of uniformity. On the more sheltered slopes the most likely colonist is hawthorn, *Crataegus monogyna* (Fig. 12.5). In gulleys protected from the wind which have a deeper layer of soil, dense thickets including Scot's pine, *Pinus sylvestris* (Fig. 8.3 p. 155), mountain ash, *Sorbus aucuparia*, and birch, *Betula verrucosa* may develop. But as we have seen, the dominant plants elsewhere are ling and heather, interspersed with areas of bracken, *Pteridium aquilinum*, and clumps of gorse, *Ulex europaeus*, or *U. galii*. In moister conditions, bilberry, *Vaccinium myrtillus*, may also predominate over considerable areas. Dispersed among the larger plants are many less obvious species such as heath milkwort, *Polygala serpyllifolia*, lousewort, *Pedicularis sylvatica*, tormentil, *Potentilla erecta*, and bird'sfoot trefoil, *Lotus corniculatus* [12.2, 12.3].

In many areas, heathland provides a low-grade form of grazing, mainly for sheep. Where this is intensive, and also in places subject to human trampling, the climax vegetation is replaced by a variety of grasses, which may form considerable swards. These grasses include the various bents such as *Agrostis stolonifera* and *A. tenuis*, wavy hair grass, *Deschampsia flexuosa*, and sheep's fescue, *Festuca ovina*. One of the most adaptable of all grass species (grazed particularly by cattle) is purple moor grass, *Molinia caerulea*, which can tolerate both dry and wet conditions.

Fig. 12.5 Sparse colonisation of heathland by hawthorn, *Crataegus monogyna*. The plant cover is predominantly ling and bell heather

Apart from the obvious adverse physical conditions associated with the heathland environment such as wind, rain, and variations in temperature, distribution and density of plant populations is influenced by two main factors: the intermittent availability of moisture and a shortage of mineral nutrients, particularly nitrate.

Water availability

While all heaths are subject to relatively heavy rainfall, the porous soils on which they occur lead to rapid drainage. The surface layer of peat possesses a marked capacity for water retention but, nonetheless, there are considerable periods during summer when xerophytic (drought) conditions exist. Among the colonists of heathland, some exhibit structural (xeromorphic) characteristics comparable with those found in sand dune communities, others make use of physiological or behavioural adaptations.

(a) Use of shelter

Many small species such as bird'sfoot trefoil, *Lotus corniculatus*, and milkwort, *Polygala serpyllifolia*, are so submerged in the surrounding ling and heather as to be almost invisible from above. In such circumstances, the microclimate differs markedly from the macroclimate a metre or so above, particularly the atmospheric humidity.

(b) Resistance to desiccation

Many species of lower plants such as mosses and lichens are able to tolerate considerable water loss during periods of drought, reducing their metabolic rate accordingly. The bristle-pointed hair moss, *Polytrichum piliferum*, is a typical example, being capable of recovery even though it has turned quite brown. Incidentally, it is one of a succession of species that readily recolonise burnt areas of heathland. Many lichens such as the various species of *Cladonia* are also a feature of heaths and these, too, are capable of recovery after becoming dry and brittle.

(c) Succulent tissue

It might be expected that a locality subject to periodic drought would be attractive to plants with water storage tissue, but evidently this is not so. One of the only examples is the English stonecrop, *Sedum anglicum* (and more rarely the wall-pepper, *S. acre*) which colonises bare areas of ground where heathers and ling are absent.

(d) Reduction of leaf surface

In gorse, the leaves have been reduced to spines, while in species such as broom, *Sarothamnus scoparius*, and petty whin, *Genista anglica*, they are scale-like, presenting only a small surface for water loss.

(e) Thick cuticle

In a number of species, water loss from the leaves and stems is reduced not only by the thickness of the cuticle, but also by its waxy covering. Typical examples are gorse, heathers, and ling.

(f) Stomata protected by folding
The folding of the leaf to form a partial cylinder with the stomata facing inwards and the outside covered with a thick cuticle is an adaptation for water conservation that we have already encountered in plants colonising sand dunes (e.g. marram grass). A similar mechanism is used by heathers on heathland.

Of the adaptations described above (a) is distributional, (b) physiological, and (c)–(f) are xeromorphic.

Nitrate deficiency

Just as an intermittent shortage of water favours colonisation by plants which can economise on its use, so the ability to supplement available mineral supplies will be at a selective advantage in heathland soil subject to leaching. A variety of mechanisms combine to achieve this.

(a) Symbiotic bacteria
The role of root nodules in nitrogen fixation was discussed in Chapter 4 (p. 60). It occurs particularly in the Papilionaceae (legumes) which are well represented on heathland. Typical examples are gorse, petty whin, and bird'sfoot trefoil.

(b) Mycorrhizal associations
This form of mutualism (p. 40) is widespread among heathland plants and is particularly noticeable in trees associated with basidiomycete fungi which form large conspicuous toadstools. Thus the fly agaric, *Amanita muscaria* (Fig. 3.9 p. 41), occurs with birch, while pines are associated with various species of the genus *Russula*. Other species forming mycorrhizal associations include the heathers, ling, bilberry, and several grasses such as the purple moor grass, *Molinia*.

(c) Insectivorous habit
This is confined to boggy areas. The sundews, *Drosera* sp. (Fig. 12.6) and butterworts, *Pinguicula* sp., have sticky, glandular hairs on their leaves which secrete proteolytic enzymes. The trapping and digestion of small insects aids an otherwise deficient supply of nitrogen. A similar habit occurs in the bladderworts, *Utricularia* sp. These are aquatic species which can trap small insects and crustaceans in little bladders where they are digested.

(d) Parasitism
This is widespread among heathland plants and varies in degree. The most extreme form, where the parasite is totally dependent upon its host, is illustrated by dodder, *Cuscuta epithymum*, which occurs as masses of pink, thread-like strands with small white flowers covering plants of heather and gorse (Fig. 12.7). The roots of several species of plants such as lousewort, *Pedicularis sylvatica*, and eyebright, *Euphrasia officinalis*, form loose parasitic associations with those of other species, particularly grasses. They are, nonetheless, capable of living an independent existence, and are therefore classed as partial parasites.

Fig. 12.6 Common sundew, *Drosera rotundifolia,* an insectivorous species (photograph Eric Hosking)

Fig. 12.7 Common dodder, *Cuscuta epithymum,* parasitising gorse, *Ulex europaeus.* The flowers of the parasite are to the right of the picture

Animals

In spite of its uniform appearance, heathland includes a greater variety of plants than is apparent at first sight. These, in turn, support a correspondingly diverse range of animal populations.

The majority of resident birds are insectivorous such as the meadow pipit, *Anthus pratensis*, and wheatear, *Oenanthe oenanthe*. Among the mammals, by far the commonest is the rabbit, whose browsing plays an important part in controlling the growth of plants such as heather and ling. Field mice (*Apodemus*) and voles (*Microtus*) are also common, as is evidenced by the frequent appearance overhead of their predator, the kestrel, *Falco tinnunculus*. Most, however, of the heathland animal community is composed of arthropods, the majority being insects. A few of these are strong flyers which are able to withstand windy conditions, for example the emperor moth, *Saturnia pavonia* (Fig. 12.8), the oak eggar, *Lasiocampa quercus*, and the fox moth, *Macrothylacia rubi*, whose larvae feed on heather. The remainder are mostly small forms which tend to shelter among the vegetation or on the ground below. These have exploited most of the plant species present. These small arthropods have been admirably described by Darlington [12.4]. Many of the herbivores are the larvae of moths. Some such as the grey scalloped bar, *Dyscia fagaria*, and the plume moth *Amblyptilia acanthodactyla*, feed on ling and heather. Others, like the larvae of the small heath butterfly, *Coenonympha pamphilus*, feed on a variety of grasses. Carnivores include the ubiquitous red ant, *Myrmica rubra*, and a large number of different beetles and spiders. Individual collection and identification of this community of heath dwelling species is not easy and the sampling of populations is best done with a sweep net.

The bog community

In low-lying areas of moorland, water tends to drain from higher ground and accumulate on the hard pan and any clay that may be present, giving rise to wet, acidic conditions. These provide an ecological environment very different from that of the heathland described above.

Plants

Lack of nitrogen is still a problem in bogs, but it is less so than on heaths because of the leaching of the soil on higher ground and the accumulation of minerals below. Nonetheless, the best known insectivorous species occur here, notably sundews, *Drosera* sp. (Fig. 12.6), butterworts, *Pinguicula* sp., and the fully aquatic bladderwort, *Utricularia*. Partial parasitism, as in heathland, is again evident in species such as the marsh lousewort, *Pedicularis palustris*. The acidity of the water in bogs is often less extreme than that on heaths (see Table 12.1 p. 238).

There is no shortage of water; rather, the problem facing plant colonists is to adapt to an excess. During periods of drought, however, bogs frequently become partially desiccated and their flora much depleted. The fact that they are usually fully restored the following season suggests a resistance to desiccation in the roots and underground stems of some species, also a durability in their seeds. Typical indicator species of boggy conditions are cotton grass, *Eriophorum angustifolium*, and bog myrtle, *Myrica gale*, which are shown in Fig. 12.9. Others include the

(a)

(b)

Fig. 12.8 Emperor moth, *Saturnia pavonia*, the largest of the heathland insects: (a) adult male (right) and female (left), (b) larva feeding on heather (photographs S. Beaufoy)

Fig. 12.9 Two indicator species of bogs; bog myrtle, *Myrica gale*, (in foreground) and cotton grass, *Eriophorum angustifolium*

cross-leaved heath, *Erica tetralix*, bog asphodel, *Narthecium ossifragum*, and toadrush, *Juncus bufonius*. Bogs are frequently carpeted with various species of bog moss, *Sphagnum* sp, which thrive in wet conditions and are able to resist desiccation. The gradual accumulation of waterlogged peat is sometimes referred to as a *blanket bog* and this is frequently colonised by *Sphagnum*, resulting in highly unstable conditions which disguise a depth of mud underneath which may extend to a metre or more. Such areas should be approached with care.

Animals

Some of the arthropods associated with bogs are similar to those on heathland, notably those herbivores relying for their food on a variety of grasses and plants such as heather. However, the peculiar flora of bogs, particularly shrubs such as the dwarf willow, *Salix herbacea*, bring with them a variety of other insect species such as the sawflies, *Pontania crassipes* and *P. dolichura*.

The mud and pools associated with bogs house a peculiar and restricted fauna. Many are burrowing forms such as Eisen's worm, *Bimastos eiseni*, and the larvae of several midges like *Chironomus* and *Tanypus*. The open water accommodates a few molluscs, notably the snail, *Columella aspersa*, and one of our largest leeches, the horse leech, *Haemopsis sanguisuga*, which feeds on a variety of invertebrates and may reach a length of 100 mm when fully extended. Other carnivores include the nymphs of several species of dragonflies, including the beautiful blue aeshna, *Aeshna caerulea*, and the northern coenagrion, *Coenagrion hastulatum*. In deeper pools it is quite common to find typical pond species such as aquatic beetles and their larvae, also the larger bugs such as *Corixa* and *Notonecta*, and the pond skater, *Gerris*. The problems of survival facing all these species are essentially similar to those in ponds as discussed in Chapter 10.

Succession on moorland

In the vicinity of a slope involving a transition from bog to heath, two processes combine to contribute to plant succession: the erosion downwards of surface soil by rain and wind, and the gradual accumulation of debris resulting from the death and partial decomposition of bog plants. As a result, the level of the ground, particularly at the periphery of bogs, tends to rise and become drier, but the rate at which this takes place and the sequence of events vary somewhat from one locality to another. The schematic outline shown in Fig. 12.10 must therefore be regarded as no more than a generalisation. For convenience, we can divide succession into three phases:

bog → wet heath → heath

but the extent of each stage is not clear cut and there is much overlap between them.

Plant species colonising bogs are required to tolerate high humidity and acidity, organic mud, and nitrogen deficiency. Typical indicator species are summarised in Fig. 12.10 and these include the two insectivorous forms, *Drosera* and *Pinguicula* referred to earlier (p. 241). The animals, as we have seen, are also mostly specific to the bog environment and include aquatic species, burrowers in mud, and herbivores, for example insect larvae which feed on the leaves of plants such as willow.

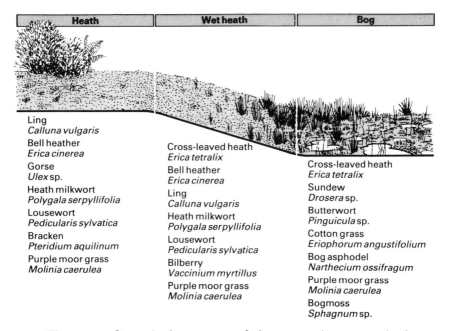

Ling
Calluna vulgaris

Bell heather
Erica cinerea

Gorse
Ulex sp.

Heath milkwort
Polygala serpyllifolia

Lousewort
Pedicularis sylvatica

Bracken
Pteridium aquilinum

Purple moor grass
Molinia caerulea

Cross-leaved heath
Erica tetralix

Bell heather
Erica cinerea

Ling
Calluna vulgaris

Heath milkwort
Polygala serpyllifolia

Lousewort
Pedicularis sylvatica

Bilberry
Vaccinium myrtillus

Purple moor grass
Molinia caerulea

Cross-leaved heath
Erica tetralix

Sundew
Drosera sp.

Butterwort
Pinguicula sp.

Cotton grass
Eriophorum angustifolium

Bog asphodel
Narthecium ossifragum

Purple moor grass
Molinia caerulea

Bogmoss
Sphagnum sp.

Fig. 12.10   Stages in the sequence of plant succession on moorland

Wet heath is the transitional zone separating bog from heath and many of its plants are similar to those in the dry area on higher ground. Some, such as bilberry, *Vaccinium myrtillus*, tend to favour damper conditions, as does the heath rush, *Juncus squarrosus*. Both species of heather, *Erica cinerea* and *E. tetralix*, are frequently found at this level.

The heath environment, as we have seen (p. 237) poses particular problems of water shortage and nitrogen deficiency, which the resident plant populations combat in a variety of ways. Many of the species commonly found on heaths are similar to those on wet heaths, but some such as bracken and gorse, prefer the drier conditions. One of the most adaptable plants is purple moor grass, which is frequently found at all three levels. The animals too, tend to be similar in both heath and wet heath, the majority being insects such as the larvae of small moths that feed on grasses and heather. These in turn support a variety of arthropod carnivores such as beetles, ants, and spiders, as well as a few bird predators like meadow pipits, wheatears, and kestrels.

Some topics for investigation

1 Survey a steeply sloping area of moorland using the bog as a datum line. Sample the plant populations using quadrats at three levels corresponding roughly to heath, wet heath, and bog. How do the plant communities at the different levels compare in terms of (a) height above datum level, (b) number of species, (c) density of species (scored in terms of relative abundance p. 86), (d) adaptations for particular ecological conditions?

2 Using the same area as in 1 above, investigate the circumstances in which *Erica tetralix* (which inhabits boggy conditions) is replaced by *Erica cinerea*.

3 Using belt transects, plot the distribution of a bog species of plant such as the sundew, *Drosera rotundifolia*. What evidence is there that soil humidity is the main factor influencing its distribution?

4 On heathland, devise a scheme for a 'unit sweep' of the vegetation with a sweep net over a given area (the size of the area will need to be determined from trial runs to see how many animals are present). Identify the catch approximately (down to class will probably do) and divide them into herbivores and carnivores as far as possible. Count the two groups and if possible weigh them to obtain the biomass of each. What can you find out about possible food relationships, and pyramids of numbers and biomass (p. 54)?

5 Repeat 4 (above) in wet heath and bog. How do the results from the three localities compare (a) in terms of diversity of species, (b) as regards biomass of herbivores and carnivores?

6 Choose an area of heath where there is a sub-dominant flora (e.g. milkwort) growing below the dominant species (e.g. ling). Make a list of the plants in each of the two groups. How do their structural characteristics compare from the point of view of resistance to desiccation?

7 As a continuation of 6 above, measure (a) temperature, (b) humidity, at ground level, just below the top of the dominant vegetation, and about 0.5 m above it. How do the two microclimates compare with the macroclimate? What relevance have these findings to those in 6?

8 Compare two areas of heathland one of which has been grazed and the other which has not. How do they differ in terms of (a) numbers of different plant species, (b) the density of each species? How consistent are your results over a

wider area? To what extent is it possible to predict the likely results of different kinds and degrees of grazing pressure (types of grazer and duration of grazing)?

9 Select an area of heathland that has been burnt (these are usually not difficult to find). What can you find out about the various successional stages involved in plant recolonisation?

Summary

1 The unique conditions existing in heaths and bogs necessitate adaptations by plant and animal colonists which are not found elsewhere.

2 Heathland soils are zoned in a particular way and are known as podsols. The surface layer of peat is low in bases and is therefore acid. Below, leaching can result in the accumulation of fine particles to form an impervious pan.

3 The heathland plant community is characteristic. Trees are frequently few or absent and are confined to sheltered areas such as gulleys.

4 The porous nature of heathland soils means water is short during the summer. Plant populations exhibit a number of characteristics which help to overcome the problem of water shortage.

5 Leaching results in a deficiency of nitrates in the surface soil. Plant colonists possess a number of mechanisms to supplement available mineral supplies.

6 Mature heathland supports a diverse range of animal populations particularly birds and insects.

7 Like heathland, bogs are also areas of low mineral concentration but they are subject to varying degrees of waterlogging. They support a typical range of plant species some of which overlap those of heathland.

8 The animals of bogs are largely arthropods and molluscs, also annelids such as worms and leeches.

9 The succession on moorland from bog to heath follows a typical pattern but the various stages are by no means clear-cut. Nonetheless, the relatively few indicator species involved makes this a good locality for study.

References

12.1 Pearsall, W. H. (1950) *Mountains and Moorlands*, New Naturalist, Collins
12.2 Gilmour, J. and Walters, M. (1954) *Wild Flowers*, New Naturalist, Collins
12.3 Turrill, W. B. (1948) *British Plant Life*, New Naturalist, Collins
12.4 Darlington, A. (1978) *Mountains and Moorlands*, Hodder and Stoughton

13 Ecological communities

Man-made habitats

In the previous five chapters we have been concerned with habitats which are natural, in so far as any communities existing today can be regarded as devoid of human interference. Here, emphasis will be on a range of habitats all of which are man-made. Some of these, such as ancient walls and hedgerows, have survived for centuries, others, like demolition areas and rubbish heaps, are ephemeral and likely to exist for only a few years. By their very nature, such sites often suffer from restricted access and a lack of diversity in their plant and animal colonists. On the other hand they may pose peculiar ecological problems which are well worth study, such as the colonisation of walls by cryptozoic animal species. Man-made habitats are often ideal for project work [13.1, 13.2].

Hedgerows

It is not easy to define hedgerows and verges precisely, for they can be a good deal more than just areas of vegetation bordering a road or intersecting a field. As Duffey has pointed out [13.3], there are about 214 858 ha of roads and verges in England and Wales of which approximately a third consist of the latter. Contrary to popular belief, verges are not just dull patches of grassland from which the weeds have been exterminated, but are frequently complex interacting communities of plants upon which many animals depend. These animals include almost half the British species of mammals and about the same proportion of indigenous butterflies. From an ecological standpoint, hedges and the verges associated with them must be regarded as a continuum.

The majority of hedgerows in Britain date from the enclosures of the 18th and 19th centuries when strip cultivation was gradually discontinued. However, some are a good deal older, even extending to Saxon times, where they sometimes provided the only means of delineating boundaries. One of the most interesting features of a study of hedges is the great diversity of plant species, particularly of woody perennials. These were either planted in the hedgerow or were incorporated from part of the pre-existing woodland. It has been estimated that the average rate of addition of new woody species per 30 m length is about one every hundred years. On this basis, a strip with ten such species could be as much as 1000 years old. Evidence derived from ancient maps and documents such as the Domesday Book suggests that such estimates are approximately correct.

Today, hedgerows serve other important functions besides marking boundaries. They may provide windbreaks and act as temporary shelter for cattle. They also have an important role as barriers to stop animals straying onto roads or from one field into another. With the advent of modern high-intensity agricultural methods, hedges have lately assumed a new importance as a sanctuary for wildlife, which would otherwise have been exterminated. Previously, meadows used for grazing would remain virtually untouched for generations, their populations of weeds being held in check by animals browsing on them. The flora of the surrounding hedgerows was closely related to that in the adjoining fields and the two merged together. Under modern practice, the ancient field has ceased to exist and leys (fields temporarily under grass) are usually ploughed about every four years, any broad-leaved weeds being eliminated by selective herbicides. Even grassland species of animals have insufficient time to colonise a newly-planted field before the habitat is destroyed once more. Thus butterflies with grass-feeding larvae, like the meadow brown, *Maniola jurtina*, which used to be common colonists of agricultural fields, are now confined to the verges round their perimeter. Although such changes are widely appreciated and documented, it is a deplorable fact that the grubbing up of hedgerows continues unabated.

Hedgerow communities

Plants

Since the essential requirement of any hedge is that it should provide a tough, inflexible barrier, it follows that woody plants must play a major part in its composition. Older hedgerows frequently incorporate large trees such as oak, *Quercus* sp and ash, *Fraxinus excelsior*, and all contain a high proportion of smaller species. These includer elder, *Sambucus nigra* (Fig. 13.1), hawthorn, *Crataegus monogyna*, blackthorn, *Prunus spinosa*, dogwood, *Cornus sanguinea*, hazel, *Corylus avellana*, and field maple, *Acer campestre*.

The range of herbaceous plants found in hedges is enormous, the principal factors determining successful colonisation being:

(i) the nature of the soil. As we have seen (p. 239), tolerance of calcium ions is particularly important in determining whether a species can survive in a lime-free (calcifuge) or alkaline (calcicole) soil.

(ii) the capacity to compete successfully with other plant species present.

(iii) the aspect of the hedge. As will be seen from Fig. 13.1, light is an important factor influencing the survival of short and long-day plants. In lesser degree, aspect also determines the degree of exposure to wind and rain, and hence the average humidity. Thus, while the north-facing community will tend to consist of flowering plants such as ivy, *Hedera helix*, and a variety of grasses such as *Holcus lanatus*, it also often includes numerous ferns such as hart's tongue, *Phyllitis scolopendrium*, and mosses like the feather moss, *Eurhynchium praelongum*. By contrast, the south-side populations are more diverse and include many typical hedgerow species such as hedge bedstraw, *Galium mollugo*, campions, *Silene* spp, speedwells, *Veronica* spp, and red dead nettle, *Lamium purpureum* (Fig. 13.2).

(iv) the policy for cutting and trimming: both the timing and frequency of the operation. This is particularly important on roadsides where cutting to improve motorists' visibility often occurs before many plants have had time to flower and set seed. Unless such species have effective means of vegetative propagation, they are unlikely to survive for long.

Animals

The animals of hedgerows are as diverse as the plants on which they depend. Earth banks provide homes for various mammals, particularly the badger, fox, rabbit, field mouse (*Apodemus*) and vole (*Microtus*). During periods of high population density, the browsing of rabbits on hedgerow vegetation may play an important part in determining its species composition. Among invertebrate

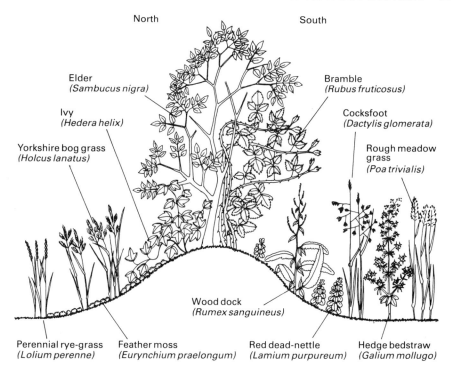

Fig. 13.1 Transect through a hedge to show variation in the north and south aspects. (See also Fig. 13.2 p. 252)

herbivores, molluscs such as the garden snail, *Helix aspersa*, and the brown- and white-lipped snails, *Cepaea nemoralis* and *C. hortensis* (Fig. 9.5 p. 170) abound. However, the majority of the invertebrates are arthropods such as the common grasshopper, *Chorthippus brunneus*, and the larvae of many moths and butterflies. Of particular interest is the gatekeeper butterfly, *Pyronia tithonus*, whose larvae feed on a variety of common grasses such as annual meadow grass, *Poa annua*. The distribution of these grasses is by no means restricted to hedges, yet the behaviour pattern of the adult insects is such that they seldom stray more than a few yards away from their hedgerow environment. Since they are easy to capture, handle, and mark, they provide ideal material for the estimation of numbers by the mark, release, and recapture technique and for studies of population dynamics.

Hedgerows as dynamic systems

Since hedges are artificial ecological entities, maintained by man for particular purposes, the natural process of succession is seldom allowed to proceed far. Associated ditches are kept clear of debris so as to drain the adjoining fields, with the result that the level of the land varies little from year to year. The introduction of the flail type of hedge cutter (Fig. 13.3) has imposed on hedgerows a more rigorous control of the pattern of plant growth than has ever been witnessed in the past. Nonetheless, the processes of inter- and intraspecific competition between different populations proceeds as intensively as ever. The numbers and density of different species are constantly changing, and as some fail to survive others take

(a)

(b)

Fig. 13.2 Well-colonised hedgerow and verge (a) south-facing aspect; the commonest species is meadowsweet (*Filipendula ulmaria*) (b) north-facing aspect; the commonest species is the common rush (*Juncus conglomeratus*)

their place. As we saw in a previous chapter (p. 233), man can exert a powerful influence in promoting such changes by transporting propagules of new species from one place to another on the surface of vehicles and by other means. Some of these changes can be quite sensational and take place remarkably quickly, for example the dense colonisation of hedgerows in parts of Wiltshire by the bluebell, *Endymion non-scriptus*, where it was almost unknown ten years before.

Fig. 13.3 Flail hedge cutter in action. The rotating head inside the cowl breaks and partially pulverises any plant material with which it comes in contact (photograph Turner International (Engineering) Ltd)

Some topics for investigation

1 Select a hedgerow for study, if possible with aspects approximately north and south. Set up a belt transect across the hedge and estimate (a) the number of different species, (b) their density, (c) the diversity index (see p. 22) for the community on each side. Could any of the differences observed between the two sides be attributed to factors other than variation in illumination?

2 Sweep a unit area of hedge on each side with a sweep net. Divide the catch of animals into herbivores and carnivores, identifying them as far as possible (not down to species). What information do the results give on (a) likely food relationships, (b) the pyramid of numbers, (c) the pyramid of biomass (if weighing facilities are available)?

3 The gatekeeper butterfly, *Pyronia tithonus,* is often abundant along hedgerows from late June to September. Estimate the size of a population by the technique of mark, release, recapture.
    (a) How do numbers vary over the period of emergence?
    (b) What variations, if any, occur in the sex ratio?
    (c) To what extent should populations occupying adjacent hedges be regarded as separate breeding entities (demes)?
  What are the implications of your findings for the ecology of the butterfly?

4 Select a series of 30 metre strips of hedge and estimate the average number of woody plant species present per strip. On the assumption that each species represents one hundred years, determine the approximate age of the hedge. Use any other data that may be available (e.g. old maps and records) to check the likely accuracy of your estimate. Is there any relationship between the age of the hedge and the diversity index of the plant community (see 1 above)?

Test your hypothesis by comparing two (or more) hedges of different ages judged by the woody plant parameter.

5 Within a particular area there are often several different methods of hedge control and maintenance, both in terms of the techniques used and their timing. What is the relationship between the various methods and the ecology of the hedgerows concerned?

Walls

Like hedges, walls are essentially man-made barriers which either form parts of buildings or act as boundaries to fields, gardens, and other property. For the study of their ecology, boundary walls are preferable, since these are usually quite low and therefore accessible, and it is frequently possible to approach both sides. An idea can thus be gained of the influence of aspect on the plant and animal colonists.

Walls are constructed of many different materials but all share the property of being strongly resistant to plant colonisation when they are new. Wooden structures are usually treated with fungicides such as creosote, while stone and brick are held together by mortar. It is not until a degree of deterioration has occurred, giving rise to small crevices where moisture and humus can accumulate, that inroads by plants and animals can begin. In some parts of Britain the walls bordering fields and gardens are 'dry', that is to say the stones of which they are composed are balanced upon one another with little or no mortar holding them together. As they age these can house diverse populations of plants and animals, but in their early years, particularly if exposed to the full force of wind and rain, they may remain almost devoid of discernible plant life. The most extensive ecological study of walls available is that of Darlington [13.4], while Bishop [13.5] has drawn up a useful list of typically mural plant species.

Three main factors influence the colonisation of walls by plants (i) the age of the wall, (ii) the material of which it is made, (iii) its aspect.

(i) As a wall deteriorates, the number of microhabitats it provides steadily increases. This is particularly true of wooden structures where breakdown leads to the formation of humus and areas of high humidity and mineral content. But even the ubiquitous lichens such as *Lecanora dispersa* which forms the familiar whitish crust on stone surfaces, require some preliminary pitting of the medium before they can establish themselves successfully.

(ii) The composition of a wall determines the nature of the habitats it provides. This applies not only to localised accumulations of humus and eroded rock, but also to the pH of the water, which will determine if the physical conditions are suitable for calcicole or calcifuge species. Thus, on limestone walls in Wiltshire where the pH is around 8.5, there is extensive colonisation by the wall-pepper, *Sedum acre* (Fig. 13.4(a)), a typical calcicole species. On old brick walls in the same area, where the pH of decomposing mortar is slightly lower, the wall-pepper is often replaced by the white stonecrop, *Sedum album* (Fig. 13.4(b)).

(iii) The most important factor determining the nature of wall communities is a continuing supply of moisture. For this reason, the north-facing sides of walls are usually more heavily populated than the south-facing sides as they are less liable to extremes of temperature and desiccation. Much of the water retained by walls from rainstorms drains from above and accumulates lower

(a)

(b)

Fig. 13.4 Two calcicole colonists of walls (a) wall-pepper, *Sedum acre*, on an old limestone wall, (b) white stonecrop, *S. album*, on a derelict brick wall

down, supporting a dense flora of species such as the dandelion, *Taraxacum officinale*, at their base. As Darlington has pointed out [13.4], run-off can lead to a vertical zonation of species in relation to water requirements such as that illustrated for three moss species in Fig. 13.5. While the upper zone of the wall remained dry and uncolonised, there was below it an area inhabited by the common moss, *Tortella tortuosa*. About mid-way down was a zone colonised by the greater matted thread-moss, *Bryum capillare*, while the foot of the wall, the region of maximum humidity, was occupied by the silvery thread-moss, *Bryum argenteum*.

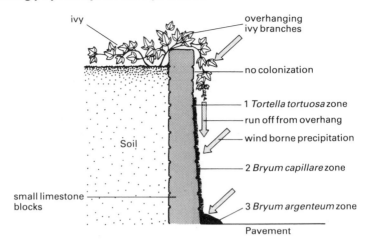

Fig. 13.5 Vertical zonation of three species of moss in relation to run-off on a wall (after Darlington)

Plant succession on walls

A vast number of different plants have been found growing on walls, ranging from lichens and mosses to woody perennials such as yew, *Taxus baccata*, and elder, *Sambucus nigra*. Some colonists are typical wall species, for example ivy-leaved toadflax, *Cymbalaria muralis* (Fig. 13.6), maidenhair spleenwort, *Asplenium trichomanes*, and wall speedwell, *Veronica arvensis*. Others like dyer's rocket, *Reseda luteola*, and Oxford ragwort, *Senecio squalidus*, are less common. Owing to the fluctuating conditions, the ecological environment of walls seldom remains stable for long, with the result that a successional sequence can be interrupted either temporarily or permanently at any stage.

These successional changes depend not only on the nature of the physical environment but also on the existence of a pioneer vegetation. This provides the starting point for further exploitation by accumulating necessities such as humus and mineral salts for use by subsequent arrivals. A theoretical succession could be:

$$
\begin{array}{c}
\text{algae} \\
\searrow \\
\text{mosses} \rightarrow \text{ferns} \rightarrow \text{annuals} \rightarrow \begin{array}{c}\text{herbaceous}\\\text{perennials}\end{array} \rightarrow \begin{array}{c}\text{woody}\\\text{perennials}\end{array} \\
\nearrow \\
\text{lichens}
\end{array}
$$

However, for the reasons outlined above, such a complete sequence is seldom achieved in practice.

Animals on walls

Animals colonise walls for two main reasons, (i) to obtain a supply of food, (ii) because the physical environment provides conditions essential to one or more phases of their life cycle. On their north-facing side, walls frequently support a

(a)

(b)

Fig. 13.6 (a) Ivy-leaved toadflax, *Cymbalaria muralis*, (b) maidenhair spleenwort, *Asplenium trichomanes*; two common colonists of damp walls

luxuriant growth of unicellular algae (Pleurococcoids) and these provide food for several mollusc species, notably the garden snail, *Helix aspersa* and the white-lipped snail, *Cepaea hortensis*. In the south, a common perennial plant colonist of walls is the red valerian, *Centranthus ruber*, which often provides a harbour for snails, particularly *Cepaea*.

Some of the most persistent plant colonists of walls are mosses, and, as might be expected, these support insect populations such as the larvae of the small moths,

Fig. 13.7 Woodlouse, *Oniscus asellus*, an inhabitant of crevices in walls in the south of England. Note the recently cast skin (photograph S. Beaufoy)

Crambus falsellus and *Scoparia* sp. But, for ecological studies, perhaps the most interesting populations of all are those of cryptozoic species, particularly woodlice. As Darlington has explained [13.4], of the five species commonly found in wall crevices, *Oniscus asellus* (Fig. 13.7) and *Trichoniscus pusillus*, are the most sensitive to environmental variations in humidity. These frequently undergo seasonal vertical migration in order to avoid desiccation: the drier the conditions the lower down is their habitat. The remainder, notably *Porcellio scaber*, *Philoscia muscorum*, and *Armadillidium vulgare* are less affected by changes in physical conditions.

Woodlice are essentially herbivores (much of their food is detritus) as are the great majority of animals living temporarily or permanently on walls. Carnivorous predators are comparatively few, the largest arthropods being the centipede, *Lithobius forficatus*, and the spider, *Dysdera erythrina* (Fig. 13.8). Where snails are

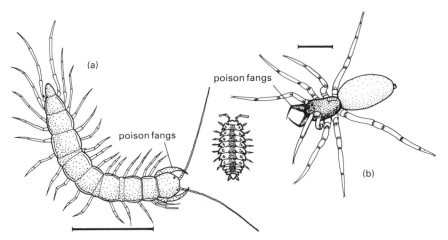

Fig. 13.8 Two predators on woodlice in walls (a) centipede, *Lithobius* sp, (b) spider, *Dysdera erythrina* (after Darlington)

common they can provide an important item of diet for song thrushes, *Turdus philomelos*, as is evidenced from anvil stones in gardens and elsewhere (Fig. 9.4 p. 170).

Some topics for investigation

1 Compare the colonisation of a wall by plants on its two aspects (preferably north- and south-facing). Is there evidence of the influence of factors other than moisture?

2 Investigate the vertical distribution of moss species (or other plants such as lichens) on a wall (see Fig. 13.5). Extract samples of the accumulated soil at different levels and estimate (a) water content, (b) organic material. To what extent can spatial distribution be related to such adaphic factors?

3 Make a study of the distribution of woodlice at different levels on walls. How are the various species present distributed in terms of (a) height of habitat above ground level, (b) humidity of atmosphere in their microhabitats (estimated by cobalt thiocyanate method, p. 109), (c) the organic content of the accumulated soil? Check your results against those obtained with a choice-chamber in the laboratory [13.6].

4 Using a marking technique (e.g. dots of cellulose paint), study the extent of movement in woodlouse populations, (a) by day, (b) by night. How do you account for your findings?

5 A wall could be regarded as providing a xerophytic environment for plant colonists. What evidence is there to support this view? For instance, how many species exhibit xeromorphic characteristics? If these are few or absent, how could this be explained?

Playing fields and paths

The essential requirement of a playing field is that it should provide a uniform and reasonably weed-free sward of grass. The species used for this purpose vary somewhat with circumstances but include bent, *Agrostis* sp, fescue, *Festuca* sp, meadow grass, *Poa* sp, and rye-grass, *Lolium perenne*. Unless stringent pre-cautions are taken (as on bowling greens and cricket pitches), invasion by a host of broad leaved weeds will take place. These weeds have been well described by Salisbury [13.7]. Some of them are herbaceous perennials and are difficult to eradicate, for example the dandelion, *Taraxacum officinale*, and dock, *Rumex* sp, which have deeply-growing tap roots (Fig. 13.9). Others include the two plantains, *Plantago major* and *P. lanceolata*, creeping buttercup, *Ranunculus repens*, clover, *Trifolium repens* and self-heal, *Prunella vulgaris*. Some weeds are annuals and therefore more fluctuating in their appearance, for example shepherd's purse, *Capsella bursa-pastoris*, pineapple weed, *Matricaria matricarioides*, pearlwort, *Sagina apetala*, and groundsel, *Senecio vulgaris*.

Playing field vegetation is subject to two main pressures, mowing and trampling.

The effects of mowing

One of the effects of regular mowing is that ephemeral species such as groundsel and shepherd's purse seldom succeed in flowering and forming seeds. Their

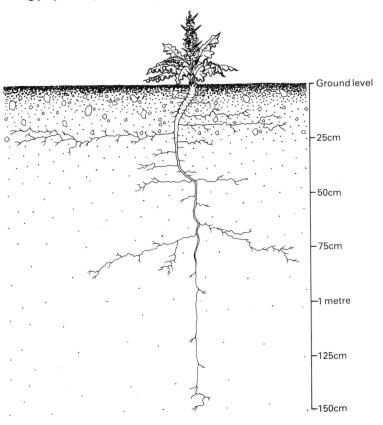

Fig. 13.9 Root system of the curled dock, *Rumex crispus* (after Salisbury)

appearance in playing fields therefore tends to be fluctuating and their density and distribution vary from year to year. The rosette habit of growth (Fig. 13.10) has powerful survival value in enabling some species (mostly perennials) to avoid the mower. This is found in dandelions, plantains, daisies, and the creeping buttercup, all of which are prevalent weeds. The widespread use of rotary mowing machines has resulted in a tendency to leave the mowings on the ground rather than removing them to a rubbish heap. This has the dual advantage of saving labour and of returning cut vegetation to the soil, thus reducing the need for additional fertiliser. Where the density of weeds such as the daisy, *Bellis perennis*, and self-heal, *Prunella vulgaris*, is high, persistent cutting can have an adverse effect. Provided conditions remain reasonably moist, the larger pieces of cut stems and leaves are capable of forming roots and thus establishing new plants. These will tend to attain their highest density in the direction of the prevailing wind.

Many playing fields are bordered by belts of vegetation which are not subject to the pressures of mowing. Here, partial succession can take place resulting in a plant (and animal) community that differs markedly from that of the adjoining sward. A comparison of the two can be well worthwhile.

(a)

(b)

Fig. 13.10 Two perennial colonists of mown grassland with a rosette habit of growth (a) daisy, *Bellis perennis*, (b) greater plantain, *Plantago major*

The effects of trampling

Heavy trampling of playing fields (e.g. for football) takes place during the winter when annual weeds have already died down. Most perennials, by virtue of their extensive underground stems and roots, are little affected. In the warmer conditions of spring and early summer, bare patches of earth are soon recolonised.

A more interesting situation occurs when a regular pathway crosses a playing field, usually along its edge. This can result, particularly during summer, in marked contrasts between the two areas in terms of plants which are tolerant to trampling such as docks, dandelions, and plantains, and those which are not (mostly annuals). The edge of the path is an intermediate zone containing some species that are trampling-tolerant and others that are less so.

The effects of herbicides

Chemical herbicides are of two kinds, non-selective and selective. The non-selective, as their name implies, are universal in their action, killing a wide variety of different plants. They are seldom used on playing fields except as a prelude to replanting and include such substances as paraquat and glyphosate.

Selective herbicides (frequently referred to incorrectly as weedkillers) are less lethal and include M.C.P.A. (e.g. 'Verdone') and 2,4D (e.g. 'Lornox') both of which are effective against many broad-leaved weeds found in fields and lawns. However, their effectiveness and the optimum time of application varies considerably from one weed species to another. Thus, creeping thistle, *Carduus arvensis*, dock, *Rumex* sp, and ragwort, *Senecio jacobaea*, are found to be most susceptible at the flowering stage, while dandelions are more readily killed in the autumn.

With the persistent use of selective herbicides, we would expect resistant strains of plants to evolve as has already occurred in several species of insects. Moreover, the elimination of susceptible species of weeds permits the spread of more resistant ones. Salisbury [13.7] records a situation in California where treatment of roadside verges killed the susceptible plants but promoted a vast population of umbelliferous weeds which proceeded to flourish in the absence of the competition that had previously existed.

Some topics for investigation

1 Investigate the density and distribution of weeds in a playing field (expressed as percentage cover) using quadrats or point frames. To make identification easier, it may be preferable to select, say, five common species. Is there a relationship between the pattern of weed distribution and (a) humidity of the soil, (b) direction of the prevailing wind, (c) any other factors such as the pattern and frequency of mowing?

2 In 1 above, what proportion of the weeds present have a rosette pattern of growth? How many are perennials? Dig up a selection of weeds and compare their root systems. How do these relate to (a) their life cycle (annual or perennial), (b) their distribution?

3 Compare a mown area of playing field with an unmown one nearby. How do the two differ in terms of (a) number of weed species, (b) proportion of rosette-forming plants? Does the uncut area show evidence of succession? If so, what are the limiting factors likely to be?

4 Investigate the distribution of broad-leaved plants in a path crossing a playing field. Is there any evidence that some species are more tolerant to trampling than others (can you devise an index of 'trampleability')? How does the situation on the path compare with (a) the field nearby, (b) the transitional zone at the edge of the path?

5 Select an area of field which has been treated with a selective herbicide, (the nature of the herbicide and time of treatment should be known). How effective

has treatment been? Is there any evidence of the resistance of certain weeds? If possible, set up a trial plot (with appropriate controls) to test your hypotheses relating to (a) resistance to the herbicide, (b) the effects of varying the time of application.

Waste land

From an ecological standpoint, waste land is a variable mixture ranging from rubbish tips and slag heaps to building sites and demolition areas. However, no matter what they are, such localities are likely to have certain characteristics in common:

(i) they are only temporary, their period of existence often being limited to a few years or less;
(ii) their ecological environment is usually diverse exhibiting considerable variations in physical conditions such as light and humidity;
(iii) because of their limited span of existence, plant and animal populations on waste areas often develop little beyond the pioneer stage;
(iv) the artificial nature of waste land means that the relationships within and between populations tend to be unstable and fluctuating.

Plant populations on waste land

The requirements of pioneer plant colonists of waste ground are essentially those of all weeds, namely a capacity for surviving in adverse edaphic conditions, particularly drought, and a high reproductive potential. Bishop [13.5] has listed a number of such species. It is not surprising to find that many are annuals, for example, shepherd's purse, *Capsella bursa-pastoris*, hairy bitter cress, *Cardamine hirsuta*, pineapple weed, *Matricaria matricarioides*, and groundsel, *Senecio vulgaris*. Herbaceous perennials which flourish in such conditions include species which propagate rapidly by underground rhizomes such as ground elder, *Aegopodium podagrarius*, couch grass, *Agropyron repens*, and self-heal, *Prunella vulgaris*; also those with creeping overground stems like silverweed, *Potentilla anserina* and creeping cinquefoil, *P. reptans* (Fig. 13.11).

In the peculiar conditions of derelict areas, competition between different plant species for the same ecological niche is often less intense than elsewhere, with the result that successional stages may follow one another rapidly or even take place simultaneously. Figure 13.12 shows a derelict building site three years after demolition. The herbaceous perennials in the foreground are the soft rush, *Juncus effusus*, indicating moist conditions, while the tree immediately behind is goat willow, *Salix caprea*, already two metres high.

Animal populations on waste land

Soil conditions on waste ground can vary greatly, even over quite short distances. Some areas such as building sites can be so desiccated and impoverished of humus as to be almost devoid of animal life. But places occupied by dumps of, say, kitchen refuse, with a high organic content and humidity can support flourishing populations of soil animals which are highly sensitive to edaphic changes, particularly desiccation. Darlington [13.8] studied such a site. The area was

Fig. 13.11 Creeping cinquefoil, *Potentilla reptans*, colonising a derelict concrete area. Note the creeping stems in the foreground

Fig. 13.12 Derelict building site three years after demolition. The herbaceous perennial in the foreground is the soft rush, *Juncus effusus*. The tree behind is goat willow, *Salix caprea*, already two metres high

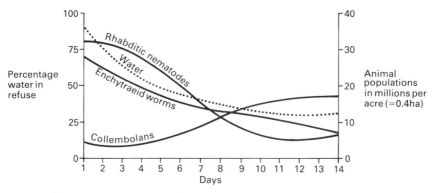

Fig. 13.13 Changes in the density of three populations of small animals in an area of kitchen refuse subject to variation in water content (after Darlington)

approximately 77 cm², and he sampled to a depth of 10 cm, extracting the organisms present by Baerman and Tullgren funnels (p. 84). His results for three groups of animals are summarised in Fig. 13.13. It will be seen that while the nematodes and worms (annelids) were affected adversely by a reduction in humidity, the collembolans (insects) benefited from such changes, which appreciably improved their environment.

Discarded objects on waste land often include pieces of timber, bricks, and lumps of concrete, which provide habitats for a range of cryptozoic animals such as woodlice and millipedes (herbivores), centipedes, beetles, ants, and spiders

Fig. 13.14 Gregarious larvae of the small tortoiseshell butterfly, *Aglais urticae*, feeding on stinging nettle, *Urtica dioica* (photograph Frank W. Lane)

(carnivores). By day, these communities remain more or less inert, but at night, when the temperature is lower and humidity high, considerable movement takes place between them as can easily be demonstrated by marking.

Many of the animals found on waste land are herbivores associated with particular plant species which may succeed in establishing themselves for a time and are then replaced by others. In Chapter 1 (p. 10) we saw how the rosebay willow herb, *Epilobium angustifolium* (Fig. 1.3 p. 8), a frequent colonist of burnt ground, may support considerable populations of the small elephant hawk moth, *Deilephila porcellus*, whose larvae eat the leaves of the plant. Another common inhabitant of waste land is the stinging nettle, *Urtica dioica*, which provides food for gregarious colonies of the larvae of the small tortoiseshell butterfly, *Aglais urticae* (Fig. 13.14). Since the plant is regarded almost universally as a pernicious weed of farms, gardens, and roadsides, it is frequently treated with selective weedkillers. This may well account in part for the recent decline in numbers of the small tortoiseshell, which was formerly abundant.

Some topics for investigation

1 On a waste land site, what is the pattern of distribution of plant species? What are the principal factors that seem to be influencing their distribution?

2 Study the pioneer plant colonists on a patch of bare ground. What proportion are annuals? Do any perennials present show particular adaptations for this mode of life (e.g. organs for vegetative propagation)? What is the possible effect of environmental factors, particularly (a) illumination, (b) soil humidity?

3 In an area of waste land which has remained derelict for several years, is there any evidence of a pattern of plant succession and factors that may be affecting it?

4 What is the distribution of cryptozoic animal populations in the area (living under bricks, logs, etc.). Estimate the relative humidity of their microhabitats using the cobalt thiocyanate method (p. 109) and compare it with the macroclimate outside. What is the numerical relationship (and biomass, if possible) of herbivores and carnivores?

5 Using pitfall traps and a marking technique (e.g. dots of cellulose paint) investigate the movement at night of cryptozoic animal species such as woodlice. It may be an advantage to set up some artificial habitats if the natural ones prove to be too far apart. What are the distances moved and the patterns of movement of the different species? What are the implications of such movement for the stability and survival of the populations?

6 Using Tullgren funnels or flotation (p. 82), investigate the density of soil organisms such as springtails (Collembola) underneath a dump of, for example, lawn mowings or kitchen refuse. (Before sampling, part of the dump must first be scraped away to expose the earth surface). Compare the results with samples from the surrounding soil. To what extent can any differences be related to (a) the organic content, (b) the humidity of the two soils?

Summary

1 The diversity of man-made habitats ranging from ancient walls to modern building sites provide exceptional opportunities for the study of certain ecological principles.

2 Besides marking boundaries, acting as barriers, and providing shelter, hedges are important reservoirs of plant and animal life.

3 The plants of hedgerows form a diverse range of communities in which trees play an important part.

4 The animals occupying hedgerows are also varied and numerous, and include mammals, birds, molluscs, and arthropods.

5 Hedgerows are dynamic systems in a constant state of change. Human activity plays an important part in influencing these changes.

6 Walls, like hedges, are essentially man-made barriers. As they age they become increasingly attractive to plant and animal colonists. From the point of view of study, they usually have the advantage of accessibility.

7 Moisture is the most important factor determining the colonisation of walls by plants and the distribution of these plants.

8 Plant succession on walls frequently follows a characteristic and predictable sequence of stages.

9 The main animal colonists of walls tend to be molluscs, insects, and crustaceans (woodlice).

10 Playing field vegetation is subject to two main pressures, those of mowing and trampling.

11 Continual mowing results in the rosette habit of plant growth having high survival value. Mowing machines can provide an effective means of spreading weeds such as daisies and the self-heal.

12 Persistent trampling along the same track can have profound effects on plant distribution. Deep rooted species tend to be tolerant while annuals are quickly killed.

13 The use of selective herbicides can have rapid effects, eliminating some plants and promoting the growth of others. This is an area of biotechnology well worth study, for instance as project work.

14 Waste land provides a wide range of different ecological environments. They are usually only temporary and relationships within them are unstable and fluctuating.

15 Derelict areas support a diversity of plants displaying the characteristics associated with weeds. Discarded objects such as pieces of timber, fallen stonework, and bricks frequently house large communities of cryptozoic animals beneath them.

References

13.1 Dowdeswell, W. H. (1981) *Teaching and Learning Biology*, Heinemann

13.2 Smith, D. (1983) *Urban Ecology*, George Allen & Unwin

13.3 Duffey, E. (1974) *Nature Reserves and Wild Life*, Heinemann

13.4 Darlington, A. (1981) *Ecology of Walls*, Heinemann

13.5 Bishop, O. N. (1973) *Natural Communities*, John Murray

13.6 Nuffield A-level Biological Science, (1970) *Organisms and Populations*, Laboratory guide, Penguin

13.7 Salisbury, Sir E. (1961) *Weeds and Aliens*, New Naturalist, Collins

13.8 Darlington, A. (1969) *Ecology of Refuse Tips*, Heinemann

14

Ecology and man

It is doubtful if any ecological environment existing today can be regarded as 'natural' in the sense that it is unaffected by the activities of man. Even in remote areas such as Alaska there is now evidence that pollutants released into the atmosphere elsewhere have been transported many thousands of miles and subsequently precipitated onto the land by the action of wind and rain. What effect, if any, such substances exert on the wild communities of plants and animals is still uncertain. But, regrettably, all too often the effects of human activities on existing ecosystems are apparent, as in the destruction of grassland to make motorways and the felling of forests for short-term agricultural gain. As we have seen in the previous chapters, the quantitative balance between producers and consumers in a food web is a delicate one and the elimination of a key link can lead to the collapse of the rest. Thus, the destruction by pesticides of the small planktonic crustaceans in a lake leads not only to a failure of the carnivores that depend upon them but also to an upsurge of the small plants (algae) that they help to keep in check. Moreover, as recent experience has shown, when such changes occur they can take place far more rapidly than was at one time supposed.

Man as a polluter

The Industrial Revolution of the eighteenth and nineteenth centuries and the massive burning of coal resulted in the addition to the atmosphere of noxious gases such as sulphur dioxide and vast quantities of soot, which tended to blacken the vegetation in the vicinity of the manufacturing cities. From the 1950s onwards, the progressive introduction of smokeless fuels has resulted in a reduction of the soot component, but sulphur dioxide and other undesirable gases such as the oxides of nitrogen still persist. Some species of plants such as the lichens are particularly sensitive to sulphur dioxide and can therefore be used as reliable indicators of pollution.

Atmospheric pollution and plants

As Richardson [14.1] has pointed out, among plants in general lichens have by far the highest sensitivity to sulphur dioxide in solution, followed by mosses, conifers, and flowering plants. The principal reason for this difference no doubt derives from the fact that, having no roots, lichens have evolved a highly efficient method of absorption over their entire body surface. Not only do they absorb solutes rapidly but they also have the capacity for accumulating them to higher levels in their tissues [14.1]. Thus living specimens of the lichen, *Usnea filipendula*, suspended in a heavily polluted area were found to have accumulated 1010 ppm of sulphur; approximately six times the amount taken up by cotton wool, glass wool, or dead lichen during the same period.

Fig. 14.1 The lichen, *Parmelia saxatilis*, a common colonist of trees and walls, and an indicator of atmospheric pollution

The relationship between sulphur dioxide in the air and the sulphur content of lichens is well illustrated by data from the vicinity of Newcastle-upon-Tyne on *Parmelia saxatilis* (Fig. 14.1), a common colonist of trees and walls. These are summarised in Table 14.1.

Table 14.1 Sulphur content of the lichen, *Parmelia saxatilis* related to sulphur dioxide in the air and distance from Newcastle-upon-Tyne (after Richardson)

Distance from Newcastle (km)	SO_2 *content of air* (ppm)	*S content of lichen* (ppm)
7	0.04	2870
13	0.028	659
34	0.028	225

However, not all species of lichens respond to similar environmental conditions in the same way, as Laundon [14.2] has shown in a study of the colonisation of gravestones at Mitcham (London) by *Lecanora dispersa* and *Caloplaca heppiana*, both common colonists of the surface of concrete and limestone. His results, spanning a period of some 200 years, are shown in Fig. 14.2. The peculiar distribution of *Caloplaca* is possibly explicable on the grounds that different stages in the life cycle of a lichen are not all equally susceptible to the effects of atmospheric pollution. Thus, whereas the thalli on the older gravestones seemed perfectly healthy, colonisation after about 1900 when pollution was increasing, appears to have been much less successful. On the other hand, the density of

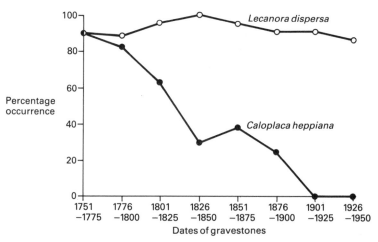

Fig. 14.2 Colonisation of gravestones at Mitcham (London) by two common lichens, *Lecanora dispersa*, and *Caloplaca heppiana* (after Laundon)

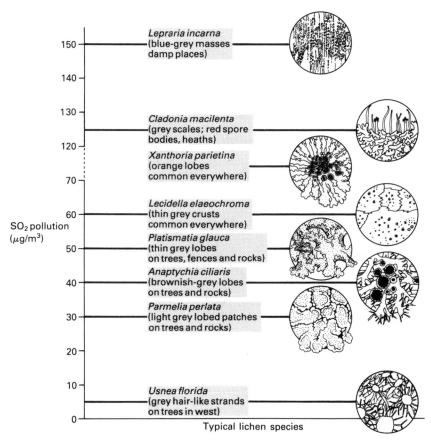

Fig. 14.3 Lichen species as indicators of sulphur dioxide pollution (data from Dalby)

Lecanora was equally high on stones of all ages, suggesting a lower sensitivity to pollutants during the early stages of growth, a supposition which was confirmed by observations on the resistance of its spores to sulphur dioxide during germination.

The differential sensitivity to pollution of different lichen species can be used as a reliable indicator of sulphur dioxide pollution of the atmosphere. Thus Dalby [14.3] has drawn up a chart, published by the British Museum (Natural History), relating varying levels of sulphur dioxide pollution to characteristic lichens. Some data from this are summarised in Fig. 14.3. At the lower end of the scale are species such as the hair-like *Usnea florida*, which colonise woodland in the west of Britain. At the other extreme is *Lepraria incarna*, which forms blue-grey masses in damp environments subject to the highest levels of pollution.

Atmospheric pollution and animals

Atmospheric pollutants such as sulphur dioxide probably exert appreciable effects on animals such as insects but these are not easy to determine with any measure of precision. One of the best instances known, which has been investigated in some detail, is the occurrence of **melanism** in insects. This is a condition associated with industrialisation, in which environmental factors have favoured the evolution of forms which are darkly pigmented due to the presence of the substance melanin (an oxidation product of the amino acid tyrosine) [14.4]. So far, studies have been restricted to moths and ladybirds but it could well prove to be a feature of other groups as well.

The evolution of melanic moths is evidently associated with the industrialised world. Over a hundred species have been recorded in Britain alone, a selection of which are shown in Fig. 14.4. Comparable changes have been reported from the industrial areas of Europe and North America. A typical example is provided by the scalloped hazel moth, *Gonodontis bidentata*, which has a dark form, *nigra* (Fig. 14.4). A century ago, only normal pale-coloured individuals were known but today, *nigra* has almost completely replaced the typical form in the area of Manchester. The classic work on melanism in moths is due to Kettlewell [14.5] who was particularly concerned with the peppered moth, *Biston betularia* (Fig. 14.4), a species in which the spread of the black form, *carbonaria*, has been closely correlated with the rise in industrial pollution. The first black specimen was recorded in Manchester in 1850, by 1950 more than 95 per cent of the population was black. Moreover, as pollution spread into rural districts, the incidence of *carbonaria* increased accordingly.

As a result of careful observation and beautifully designed experiments, Kettlewell was able to show that insectivorous birds such as the hedge sparrow, *Prunella modularis*, the spotted flycatcher, *Muscicapa striata*, and the nuthatch, *Sitta europaea*, selected the moths differentially according to their background. Thus in unpolluted areas with lichen-covered trees, the typical light form was at an advantage, while in polluted districts with soot-blackened vegetation and a lack of lichens, *carbonaria* possessed the advantage of concealment. Kettlewell subjected these observations to quantitative test [14.6] by rearing and liberating large numbers of typical and melanic peppered moths in two widely separated localities, one in Dorset (unpolluted) and the other in Birmingham (polluted). His results are summarised in Table 14.2.

Fig. 14.4 Some species of moth exhibiting industrial melanism:
(a) peppered moth, *Biston betularia* – typical, melanic (*carbonaria*), melanic (*insularia*)
(b) scalloped hazel, *Gonodontis bidentata* – typical, melanic
(c) waved umber, *Hemerophila abruptaria* – typical, melanic
(d) nut-tree tussock, *Colocasia coryli* – typical, melanic
(e) willow beauty, *Cleora rhomboidaria* – typical, melanic (rural in England, industrial in Germany)
(f) mottled beauty, *Cleora repandata* – typical, melanic
(g) tawny-barred angle, *Semiothisa liturata* – typical, melanic
(photograph John Haywood)

Table 14.2 Numbers of the peppered moth, *Biston betularia*, liberated and recaptured in two different localities (after Kettlewell)

Locality			Typical	Carbonaria	Total
Dorset 1955 (unpolluted)	{	Liberated	496	473	969
		Recaptured	62	30	92
		% recaptured	12.5	6.3	
Birmingham 1953 (polluted)	{	Liberated	137	447	584
		Recaptured	18	123	141
		% recaptured	13.1	27.5	
Birmingham 1955	{	Liberated	64	154	218
		Recaptured	16	82	98
		% recaptured	25.0	52.3	

From the percentage recaptured it is clear that while the typical light form enjoyed an advantage in the unpolluted area, it was at a marked disadvantage in the polluted zone, where *carbonaria* predominated.

Subsequent studies have indicated that, as frequently occurs in ecological situations, a number of factors have combined to bring about the remarkable rise of melanism in *Biston betularia*. These can be summarised briefly as follows:

(i) Soot

This has undoubtedly been a major factor in providing a background on the trunks of trees against which *carbonaria* is relatively invisible to potential predators. It is significant that with the introduction of smokeless zones from the 1960s onwards, the incidence of melanics has declined appreciably in areas where they previously existed at a high density.

(ii) Sulphur dioxide

Estimates of the average sulphur dioxide concentration in industrial areas in 1958 (and no doubt before that) show it to have been around 200 μg m^{-3}. As Fig. 14.3 indicates, this is a level of pollution that no lichens can tolerate. Their absence from tree trunks must therefore have reinforced the darkening effect of soot mentioned earlier and favoured the survival of *carbonaria*. By 1970, the sulphur dioxide concentration is estimated to have dropped to around 120 μg m^{-3}, a level at which lichens such as *Lecanora* spp are able to survive. Their reappearance could have provided a background more favourable to the camouflage of the typical *Biston betularia*, reinforcing the effect of reduced soot pollution referred to above, and favouring the decline of melanics. A further point worth noting is that smoke tends to be carried by the wind at low altitude away from the area of production. We would, therefore, expect gradients of melanism down-wind from the sources of pollution. This is precisely what Kettlewell found. Concentrations of sulphur dioxide tend to be dispersed at higher altitudes and over longer distances, much of it returning to earth dissolved in rain. In areas of high sulphur dioxide concentration but low soot, we might therefore expect an absence of those lichen species which could provide protection for typical *B. betularia* at rest and a consequent rise in the melanic population. This is what seems to have happened in East Anglia where the level of *carbonaria* is over 70 per cent.

(iii) Differential survival of larvae

Kettlewell found that in some broods of industrial origin, the caterpillars destined to give rise to pale adults tended to feed up and pupate earlier than the others, thus avoiding the gradual accumulation of sooty vegetation on which they did not survive well. Moreover, slow feeding could have allowed *carbonaria* larvae to rid themselves of any accumulation of toxic substances which might have interfered with subsequent development. This kind of advantage in a young stage has also been detected in other moth species such as the mottled beauty, *Cleora repandata* (Fig. 14.4), which also exhibits industrial melanism. Here, the gene causing blackening in the adult has been shown by Ford [14.7] to benefit the larvae as well as it confers greater hardiness and enables them to withstand starvation better than larvae of the typical form.

Other forms of pollution

The catalogue of man's polluting activities makes depressing reading and there is no point in reciting it here. As we have seen, the response of living communities to increasing pollution is either to succumb or to undergo modification under the influence of natural selection. Some of the most serious examples of pollution have occurred in our major rivers. Happily, too, their restoration has provided one of the success stories in conservation. A typical example is the River Thames [14.8], which reached its lowest point in the mid-1900s when most of London's sewage was poured into it untreated. A 42 km section of the river in Central London became so foul that it was devoid of dissolved oxygen and therefore of aerobic life. During the summer months the smell of hydrogen sulphide resulting from the anaerobic putrefaction of human excreta and other equally undesirable effluents was intolerable. Following local government and other enquiries initiated in 1951, a new complex of establishments for the treatment of sewage and monitoring of effluents was set up. The input of water from the various industries bordering the river was also carefully checked for polluting agents. Since then, regular chemical and biological surveys have monitored the changes occurring. These have resulted in a 65 per cent reduction in pollution and a corresponding rise in the oxygen content of the water. Even at London Bridge the concentration does not fall below a level of 45 per cent saturation. In 1982, the one hundred and first species of fish was recorded, an indication of the vast increase in the aquatic plant and invertebrate communities. In the same year 30 fully grown salmon were caught at Molesey Pool, near Hampton Court. Perhaps the most impressive evidence of revival of the river community has been the granting by the Thames Authority of a commercial licence to trap eels once more after a lapse of more than 100 years, showing that the ecology of the river is now sufficiently established to withstand controlled cropping.

Pollution and agriculture

Paradoxically, some of the most powerful pollutants have been developed for agricultural purposes, where they have conferred great benefits on mankind. Thus the various organochlorine compounds such as DDT and Dieldrin were released into the soil in increasing quantities from the 1940s onwards and proved highly effective in the control of a wide range of insect pests. The disadvantage of such insecticides is that they are largely unselective and therefore tend to destroy beneficial and harmful species alike. Evidently, innate tolerance of these poisons varies from one pest to another. Where it occurs, it confers a great selective advantage on the individuals concerned and in conditions of repeated spraying resistance tends to increase, particularly among species such as aphids with a high rate of reproduction. Thus, today we have races of DDT-resistant cockroaches and flies, also aphid populations in glasshouses which are almost unaffected by several commonly used insecticides. Moreover, as growers of crops such as tomatoes have discovered, insecticide tolerance can build up rapidly in a matter of a few years.

Man as an ecological factor

As Yapp [14.9] has pointed out, few people realise that the greater part of the landscape of Europe and of North America east of the Rocky Mountains has been man-modified. Wherever he has penetrated, man has grown crops and kept domestic animals, and in so doing has destroyed old habitats and created new ones. Such exploitation has wrought vast changes in the balance of naturally occurring ecosystems.

Destruction of hedgerows

In Chapter 13, brief consideration was given to the survival of hedgerows and the problems created by their elimination. A survey made by Locke in the 1960s showed that England, Scotland, and Wales together supported some 985 600 km (616 000 miles) of hedges, representing approximately 65 per cent of all fields and other kinds of boundaries [14.10]. However, the economic use of modern machinery such as large combine harvesters depends upon the existence of uninterrupted tracts of cultivation and this has led to the removal of hedgerows at an alarming rate. The effects of removing hedges (and other long-established ecological habitats) can be both direct and indirect. Direct effects are the elimination or reduction of the number of plant and animal species previously present. Thus, an average farm of around 101.2 ha (250 acres) with small fields bounded by hedges could normally be expected to support at least forty different species of breeding birds. But on farmland denuded of hedges the bird population can be reduced to as little as three breeding species, the partridge, *Perdrix perdrix*, skylark, *Alauda arvensis*, and lapwing, *Vanellus vanellus*.

Indirect effects derive largely from recent changes in agricultural practice and the ploughing up and reseeding of leys (fields temporarily under grass) about once every four years, also the treatment of the vegetation with selective herbicides to kill all broad-leaved weeds. As a result small herbaceous perennials such as rest harrow, *Ononis spinosa* and bird'sfoot trefoil, *Lotus corniculatus*, both food plants of the larvae of the common blue butterfly, *Polyommatus icarus*, scarcely manage to establish themselves before they are destroyed either by spraying or ploughing. Before the present practice, when fields were allowed to mature, populations of common blues were common; today, they are virtually unknown. However, appropriate food plants and ecological conditions often still exist in hedgerows, which therefore provide a refuge for many species like the common blue that formerly enjoyed a wider distribution. Once the hedges are removed, the insects have no alternative habitat available to them and the inevitable consequence is extinction. Comparable human interference, in this case the draining of the fens, undoubtedly contributed (aided by the cupidity of collectors) to the extinction of one of our finest butterflies, the large copper, *Lycaena dispar*, the last specimens being captured in about 1848.

Planting of conifers

Just as man has been a destroyer of ancient habitats, so he has been a creator of new ones. The concept of the monocrop, the cultivation of a single kind of plant over a wide area, is fundamental to modern agricultural practice. The one crop can assume many forms such as grass, cereals, root crops, and crucifers like rape, *Brassica napus*. Where land is deemed unsuitable for the cultivation of such crops, it has been turned over to the Forestry Commission for the establishment of woodland which consists predominantly of conifers such as our native Scots pine, *Pinus sylvestris*, larch, *Larix decidua*, and Sitka spruce, *Picea sitchensis*. Today such plantations have become a feature of many of our hillsides (Fig. 14.5), particularly in the north of Britain.

The choice of conifers for cultivation in preference to broad-leaved trees must be judged, at least in part, against an economic background. Here, the following factors are important:

(i) the availability of land
Where areas are designated for afforestation they tend to be on leached soils which are better suited to the cultivation of conifers than of broad-leaved trees.

(ii) national needs
Taking into account our requirements for sawn timber, poles, and paper pulp, Edlin [14.11] has estimated that our softwood needs are about six times as great as

Fig. 14.5   A typical plantation of conifers (Gwydyr Forest) (photograph Forestry Commission)

those for hardwood. However, roughly half our hardwood requirement is for tropical timbers which cannot be grown in Britain, so omitting these from the equation, we have a ratio relative to the pattern of use of one broad-leaved tree to every twelve conifers.

(iii) productivity

Comparative values calculated by the Forestry Commission for the increment of useful timber by different species of forest trees (mg ha^{-1} year^{-1}) show oak to be 4.2 which is similar to Scots pine, 4.3. By contrast, the figure for Corsican pine is 7.3, while that for Sitka spruce is 9.2 (more than double that of Scots pine). This species has been introduced from the Pacific coast of North America where it flourishes on acidic soils in conditions similar to our own.

Once a coniferous forest has passed the first few years of growth, a lack of light virtually precludes colonisation by resident herbaceous plants. These become confined to occasional clearings, and fire lanes, which are often mown, thus modifying considerably their ecological environment (Fig. 14.6). Nonetheless, they can support considerable colonies of grass-feeding species such as the grasshopper, *Psophus stridulus*, and the small heath butterfly, *Coenonympha pamphilus*. As Yapp [14.9] has shown, the planting of conifers has brought considerable benefits to a number of animal species. Thus the red squirrel, *Sciurus vulgaris*, is almost confined to this kind of habitat, and the capercailzie, *Tetrao urogallus*, entirely so. The kind of rough grassland likely to be converted to fore·

Fig. 14.6 Young and older coniferous forest. Fire lanes (in foreground) and other clearings can support considerable colonies of grass-feeding species of animals (photograph Forestry Commission)

will usually support only two common bird species, the skylark, *Alauda arvensis*, and the meadow pipit, *Anthus pratensis*. But once the young trees grow together and the grass below them disappears, these are replaced by up to twenty new colonists, including such common species as the robin, *Erithacus rubecula*, the chaffinch, *Fringilla coelebs*, and the tree pipit, *Anthus trivialis*. These are mainly insectivores, and in order to maintain the new ecosystem they must be supported by a large community of insects which draw their food from different parts of the forest trees.

Finally, we should not overlook the physical effects of these new plantations, which provide huge wind breaks and make the climate downwind appreciably warmer. Moreover, the roots of the trees constitute a barrier to the run-off of water, which reduces the liability to erosion of the hillside land below. Judged aesthetically, coniferous forest may present a somewhat drab and monotonous appearance, but it has compensating advantages both for the ecological and economic benefits that it confers.

Modification of ecosystems

The two previous examples, the removal of hedgerows and the establishment of coniferous forests, both involved the elimination of established and possibly ancient ecosystems and their replacement by new plant and animal associations. In each instance the effects were drastic, for instance in reducing the number of breeding bird species from around forty to three, or introducing new forms which had not occurred there before. Unfortunately, we are still far from understanding all the interactions within even the commonest ecosystems, so it is seldom possible to predict precisely what the outcomes of particular changes in ecology are likely to be. The important point is that when making such changes we need to be vigilant, in the hope that we can correct adverse situations before they get out of hand. This is particularly true where a potentially destructive species is transferred from one environment to another, as has happened with the gipsy moth, *Lymantria dispar* (Fig. 14.7). The species used to be indigenous in Britain

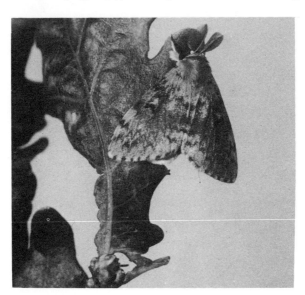

(a)

(b)

(c)

Fig. 14.7 The gipsy moth, *Lymantria dispar*, a species introduced from Europe to North America, where it has become a major pest; (a) adult male, (b) adult female, (c) larva (photographs S. Beaufoy)

and periodic specimens are still found, but these are probably escapes, as the moth is bred quite widely from foreign stock. It is, however, common in Europe and late last century a French astronomer working at Harvard University brought some eggs from Europe to his home in Massachusetts. Some of these (or the young larvae hatched from them) somehow went astray and started one of the biggest caterpillar plagues ever, attacking the forests, orchards and gardens of New England. In spite of an immense amount of research, the introduction of numerous possible parasites and predators from Europe, and the expenditure of vast sums of money, *Lymantria* still remains a major pest of a wide variety of trees, many of them of great economic importance [14.12].

Man as a consumer of resources

As we saw in Chapter 4, the recycling of materials such as essential minerals, is a feature of all natural ecosystems. In recent years there has been an increasing awareness that resources essential to human existence worldwide are finite. Moreover, following extensive research, we now have a fairly precise estimate of the extent of these resources and where they are located. The rate of consumption varies with world trade; during periods of expansion it rises and in recesssion it falls. Just how rapid increase can be in the use of a common metal such as iron is illustrated by the rise of steel production in Britain over a hundred year period, illustrated in Fig. 14.8. Thus between 1870 and 1920 output increased by a factor of 25, while for the 50 years 1920–1970, production trebled. The realisation that the sources of essential metals such as iron, copper, zinc and aluminium are becoming rapidly depleted has prompted greatly increased attention to salvage and the reuse of scrap metal.

Another aspect of salvage relates to our increasing consumption of energy and the need to explore new sources. It is estimated that the households, shops, and offices of the United Kingdom provide about 19 000 000 tonnes of waste annually, with an energy value of between 8800 and 10 000 kJ kg^{-1}. The estimated energy values of some common constituents of refuse are shown in Table 14.3.

Table 14.3 Constituents of refuse and their energy values

Constituent	% by weight (national average)	Energy value (kJ kg^{-1})
Fine dust (<20 mm)	19	9600
Paper	30	14 600
Vegetable matter	21	6700
Rags	3	16 300
Plastic	3	37 000
Unclassified (wood, leather etc.)	6	17 600

Processes have now been devised to convert refuse into solid waste-derived-fuel (WDF) which could well make an important contribution in the future, particularly to the requirements of large commercial concerns such as power stations and factories.

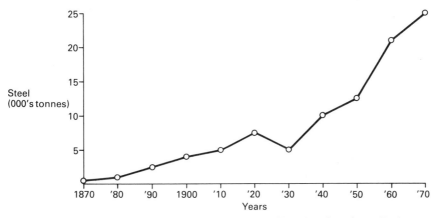

Fig. 14.8 One hundred years of steel production in Britain (data from Barker *et al.*)

Control of resource depletion

In the absence of the necessary technology to achieve complete recycling of resources, or to utilise alternative materials to those in increasingly short supply, the next best thing is to limit net consumption to the minimum possible. Cloud [14.13] has devised three alternative patterns of mineral resource depletion and predicted their consequences (see Fig. 14.9). Under present conditions characterised by unrestricted extraction and use with low recycling and high wastage, supplies of some key metals will be exhausted by the year 2000 (curve A). The depletion time could be substantially increased by partial recycling and less wasteful use, but the outcome is the same although longer delayed (curve B). High efficiency of recycling combined with stringent conservation and the increasing use of alternative materials (curve C) could lead to a greatly reduced rate of depletion where existing supplies might last almost indefinitely. However, even if we were able to achieve perfect recycling, some wastage would inevitably occur due to such factors as corrosion, friction, and loss to the soil. Thus it has been

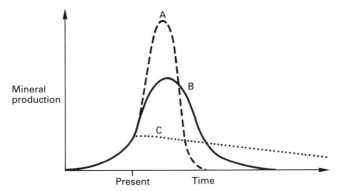

Fig. 14.9 Alternative depletion patterns for mineral reserves: A – unrestricted extraction and use, low recycling; B – partial recycling, more efficient use; C – efficient recycling and conservation, effective substitutions (after Cloud)

estimated that if we were to recycle 60 million tonnes of iron each year, about half a million would still be needed if all losses were to be made good. This only serves to underline the urgency of the search for alternative materials.

Protein as a resource

Proteins are an essential requirement for the maintenance and growth of cells. All of them are composed of some twenty amino acids, united chemically in a great variety of ways. Of these, man is capable of synthesising nine, the remainder must be obtained from other sources. Animal flesh and animal products such as eggs and cheese contain proteins which are similar to ours and, when digested, these provide the amino acids we need. These proteins are high grade in the sense that when they are eaten no chemical conversion of amino acids is required and there is therefore little waste. Proteins derived from plants lack certain of the essential amino acids and a greater amount of this protein is needed for the same effect, so they are regarded as a lower grade source.

Proteins contain, on average, 16 per cent nitrogen and this provides a useful parameter by which to assess requirements. Thus the level of protein intake needed to maintain health can be defined as the amount that will achieve nitrogen equilibrium in adults and satisfactory growth and metabolism in children [14.14]. Precisely how this is determined need not concern us here but a report by our Department of Health and Social Security gives a minimum protein requirement for an adult man of 45 g per day. The consumption of protein in different parts of the world varies enormously. Thus the Food and Agriculture Organisation of the United Nations has estimated that while the intake of meat and fish by a city dwelling American may reach 194 g per day, that of a labourer in India is 2 g per day. But the problem is not only one of vast inequalities, it is compounded by a world population increasing by approximately 2.5 per cent per year. Thus, by the year 2000, it is estimated that there will be a world shortfall of high grade protein of around 20 million tonnes.

World meat production, mainly in the form of cattle and sheep, is strongly

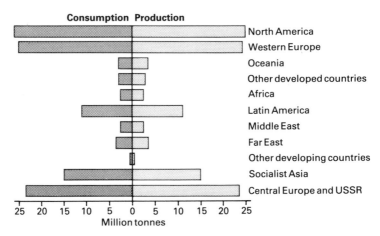

Fig. 14.10 World food production and consumption. Oceania consists of Australia, New Zealand, and the South West Pacific Islands (from BP Nutrition Ltd (1975), *A Protein Survey*, B.P. Nutrition)

localised, as is shown in Fig. 14.10. Thus, in Europe and North America climatic conditions and soil fertility lead to high productivity, in contrast to much of Africa where rainfall and production are so low that it may require as much as 40 ha (100 acres) or more to support a single bullock for a year, approximately twenty times the comparable area in Britain.

As we saw in Chapter 4, one of the characteristics of ecosystems is the loss of energy that occurs in food conversion from one trophic level to the next. At the herbivore level, the wastage may be as high as 90 per cent. But in assessing the overall food requirements of herbivorous mammals such as cattle, we must also take into account the losses occurring in vegetation such as grassland due to decomposers and other herbivores like insects and molluscs. Thus, on an average ley (field temporarily under grass), a bullock might be expected to consume roughly 15 per cent of the energy available to it. Of the remainder, as much as 70 per cent will be lost due to trampling and the subsequent breakdown of the dead grass stems and leaves by decomposers, while the remaining 15 per cent is eaten by other herbivores.

In our present state of knowledge, the most economic way of converting solar energy into the chemical energy of protein is through the flesh of herbivores (primary consumers). This, at least, avoids a further loss of energy at the secondary consumer level. Domesticated animals vary considerably in their ability to convert the protein in plant food to animal protein. Bowman [14.15] has summarised such information showing the present biological performances (the ratio of protein taken in to protein produced) for a number of farm animals, and the ceiling considered possible under optimum conditions (Table 14.4).

Table 14.4 Efficiency of protein conversion by domestic animals (after Wilson)

Species	Biological performance (%)	
	Present level	*Ceiling level*
Cow – milk	28	46
Cow – beef	6	10
Sheep	4	12
Poultry – eggs	20	33
Poultry – meat	20	31
Pig	12	18
Rabbit	11	20

Compared with the supplies of minerals considered earlier, the world protein situation is not only one in which a resource is already deficient and unequally distributed, but is one where the position is getting worse due to an increasing world population. Clearly, there is an urgent need to combat the deteriorating situation and a number of different approaches are possible.

(i) Increasing the proportion of land under cultivation
Great strides have been made in this direction in recent years, particularly through schemes of irrigation. These are expensive (e.g. the construction of dams) and economics are frequently a limiting factor. Moreover, it must be borne in mind that the proportion of the earth's surface suitable for the growing of grass is strictly limited.

(ii) Improvement in agricultural practice

Considerable progress has been made here too, with the introduction of advanced methods such as mechanisation and the production of improved strains of crops. One of the reasons for the present world shortage of foodstuffs of all kinds, not only protein, is that as the areas of grass become insufficient to support the necessary farm stock, so an increasing proportion of human food such as corn, maize, and barley is being fed to animals. As a community becomes more affluent, so the demand for animal products increases, in preference to traditional sources of protein provided by plants such as the pulses (peas and beans). The present preponderance of demand by the developed nations in comparison with that of the Third World is clearly illustrated in Fig. 14.10.

(iii) Redistribution of food stocks

As Fig. 14.10 shows, the distribution of existing world food stocks is very unequal. It has been estimated that if only the required amounts of the food available now could be transported to the right places at the right time, present needs could be met without any further increase in production. The problem is one of economics and distribution, two closely related factors in the international scene.

(iv) Changes in feeding habits

There is no doubt that protein consumption by the wealthy nations far exceeds requirements. The problems of effecting changes in feeding habits involve not only nutritional considerations but also palatability. For most people, animal products taste more pleasant than those derived from plants. In recent years, considerable advances have been made in the manufacture of more attractive plant products from species with a high protein content such as soya beans. The next step will be to gain a wider degree of acceptance.

(v) Synthetic production of protein

It may well be that in the future, at least part of the world's protein needs will be met from synthetic sources. Pilot plants such as that producing Toprina from yeast using part of the crude oil distillate as a feedstock (BP Nutrition Ltd) [14.16] have demonstrated that the large scale manufacture of protein is a practical proposition once the process can be made economic. Other possibilities include the culture of bacteria as a continuous process, also the culture of the filamentous fungus, *Fusarium*. Whether this is a justifiable use of valuable energy and chemical resources is open to debate.

Ecological technology

The contributions of ecology to our technological society are now so diverse that it would require a separate book to accord them all adequate treatment. Some are unobtrusive, while others occur in somewhat specialised contexts and tend to be overlooked by students of biology. But first let us be clear about what we mean by technology, which is not synonymous with applied science, as is commonly supposed. A technological approach involves not only the application of the methods and findings of a science such as ecology to the solution of human problems but also a consideration of the economic, social, and other implications. It is therefore a totality, aiming at the overall coverage of a problem. Among its many technological aspects, biology boasts two of the oldest technologies in the world, medicine and agriculture. Both have strong ecological affinities and therefore provide good examples of ecological technology.

Ecology and medicine: the malaria problem

Malaria has been described as the greatest killer of all time. Among its victims were Alexander the Great (336 BC), Oliver Cromwell (1658), and Lord Byron (1824) [14.17]. The fall of the Greek and Roman civilisations has been attributed in part to the devastations of malaria. Today, it is confined to the hotter parts of the world on either side of the equator, but in the Middle Ages it was common in Britain, particularly in Scotland, East Anglia, and the South. The disease can be prevented and usually cured by drugs such as pyrimethamine (Daraprim) taken orally. Traditionally, quinine was used for this purpose but is much less effective. However, prevention by prophylactics (protective drugs), while highly successful in disciplined organisations such as an army, is quite impracticable where large civilian populations are concerned. Moreover, although the disease is by no means always fatal, it leaves its victims in a much weakened state and therefore more likely to catch any other diseases that may be about.

The only hope of solving the malaria problem is to eradicate the disease at its source and it is here that ecology has made an important contribution. Information on the parasite causing malaria (*Plasmodium*) and the female anophelene mosquitoes that carry it (so called because most of them belong to the Genus *Anopheles*) can be found in most biology textbooks and will not be repeated here. Suffice it to add that the eggs, larvae, and pupae of the insect are all aquatic. The adult males are harmless and live on plant juices, while the females have well-developed piercing and sucking mouthparts and feed on the blood of a variety of mammals, including man. Since *Plasmodium* is a parasite of red blood cells, the mosquito provides an effective means of transmission from one human being to another (i.e. it acts as a **vector**).

In 1955, the World Health Organisation decided to implement a programme for the world-wide eradication of malaria. This was to be done in three ways:

(*i*) *Draining the breeding grounds*
This can be highly successful provided conditions are right. Thus, draining the marshes near Rome completely eliminated malaria and freed 8000 ha of fertile farmland for cultivation.

(*ii*) *Destroying the larvae and pupae*
Spraying with light oil mixed with poisonous substances and the introduction of predatory species of fish such as the genus *Gambusia*, can be effective means of controlling larvae and pupae.

(*iii*) *Killing the adults in their resting places*
The insect flies at night and rests by day. The effective use of insecticides thus depends on a knowledge of its habits and the use of chemicals to which it is susceptible.

The effective application of the three methods of control have depended on a detailed knowledge of the ecology of the mosquito and its variations in different species and localities. While methods (i) and (ii) have proved successful where circumstances have permitted, eradication has been concentrated mainly on the spraying of adult mosquitoes with modern insecticides such as DDT, benzene hexachloride (BHC), and dieldrin, which can remain active on the walls of houses for several months.

Some of the results achieved through these programmes of concentrated spraying have been sensational, a typical situation being that in Sir Lanka between 1946 and 1971 [14.18], which is summarised in Table 14.5.

Table 14.5 Control of malaria in Sri Lanka, 1946–71 (after Service)

Year	Cases of malaria reported	Control measures
1946	2 800 000	None
1961	110	DDT spraying
1964	150	Spraying ceased
1966	499	Spraying ceased
1968	440 644	Spraying ceased
1969/70	1 500 000	Spraying ceased
1971	145 368	Spraying renewed

The correlation between the incidence of spraying and the level of malaria is striking. But the main factor determining the success of such programmes is economic. Can a country afford to do the job properly? A second factor that has recently come to light is one we have already encountered (p. 275), namely the evolution of strains of insects resistant to certain insecticides following repeated spraying. This has certainly occurred in Sri Lanka, where the continued use of DDT has led to a decreasing sensitivity of adult mosquitoes in certain areas. This serves to underline the importance of developing alternative forms of spray, also of bearing in mind the possibility of developing more efficient methods of killing the young stages in water.

Ecology and agriculture: the problem of fungal disease of cereals

Ever since man began cultivating crops he has had to contend with a large array of pests, including those that ravage stored products and parasites which reduce yields. Among the latter are a range of fungi traditionally referred to as smuts and bunts, one of the commonest in temperate zones being *Erysiphe graminis* which causes powdery mildew on the leaves of wheat, barley, and oats, *Erysiphe* exists in a number of different strains, each specific to one kind of cereal, so it does not spread from one crop to another. The fungus at first forms whitish fluffy patches on the lower leaves, but as the mycelium spreads, the whole plant becomes infected, the leaves turn yellow and shrivel, and the root system fails to develop completely. Spores are produced in vast numbers on the leaf surfaces and these are quickly spread by the wind and infect other plants. Once the disease has gained a foothold, a whole field of young wheat can turn yellow in a matter of a few weeks. The spores are highly resistant to adverse conditions such as desiccation and low temperatures and provide an effective overwintering stage in the soil. Should a similar crop be planted the following year it will quickly become infected.

The effects of these fungal attacks on cereals vary greatly. Mild infections may result in as little as 5 per cent drop in yield; severe attacks can be devastating, causing losses of up to 50 per cent or even total destruction. The need to combat fungal disease has been recognised by farmers for many centuries, although it is only comparatively recently that the life cycles of the various pests have been worked out. Several methods of control have been evolved which can be summarised as follows:

(i) Cultural control
This is the traditional procedure based on the principle of rotation and avoiding the growth of a susceptible crop on ground previously occupied by an infected one. Thus plants remaining in a field after harvest are destroyed and, when possible, stubble is burnt in order to reduce sources of infection. The limitations of this method of control are that destruction of residual spores is far from complete and considerable areas of reserve land are needed, as well as flexibility in crop production, in order to achieve a measure of success.

(ii) Use of resistant strains
Despite the efforts of plant breeders, the production of varieties of cereals resistant to fungal infection has been only partially successful. No cultivar (cultivated variety) can be relied upon to be totally immune to infection. The best that has been achieved so far are **field tolerant** strains which, although susceptible to some extent, do not suffer as much damage as non-resistant forms. In rapidly reproducing species such as fungi, it is not surprising to find that mutant varieties have arisen which have the capacity to overcome host resistance, and are resistant to the effects of some chemical fungicides.

(iii) Treatment of seeds
It has long been known that fungal infection can be carried in or on the seed. Thus Jared Eliot, writing in 1748 says, 'If a farmer in England should sow his wheat dry without steeping in some proper liquor he would be accounted a bad husband-man'. The liquor traditionally used was salt solution, its concentration being tested by using an egg as a hydrometer. Various other treatments followed, including the use of copper sulphate, methanal, and a range of mercury compounds with varying degrees of toxicity. Today, these have been superseded by the systemic fungicides described in the next section.

(iv) Systemic fungicides
It is not surprising that it has taken much longer to produce systemic herbicides than it has to develop systemic insecticides. The problem is to find chemicals which can be absorbed by the plant without harm but which poison any fungus feeding on it. The first of the modern range of substances (triamiphos) was introduced in 1960. It was followed by many others with a wide spectrum of activity, examples being benomyl and thiabendazole, both widely used in gardens. They are not toxic to human beings and on reaching the ground are quickly broken down.

Today, the spraying of a field of young wheat infected with *Erysiphe* achieves a remarkable transformation from the yellow of infection to the green of heatlh in a matter of a few weeks. This has been achieved through ecological technology – the application of ecological principles and chemical research, with due consideration of economics both in production and in application.

Nature conservation

In a previous section (pp. 280–4) we were concerned with the conservation of the world's dwindling physical resources such as minerals. Here, the emphasis will be on our organic resources – the plant and animal communities that together make up the biotic environment of Britain. As we have seen in the earlier chapters, ecology is concerned with three broad issues (i) the nature and functioning of ecosystems, (ii) changes occurring in the structure and density of populations, (iii) the adaptations and reactions of plants and animals to different physical and biotic conditions. The process of **conservation** implies the maintenance of a certain predetermined set of conditions. The outcome is frequently the *preservation* of a habitat, community or population. The two processes, although sometimes similar in their aims, must not be confused. The interests of the conservationist are similar to those of the ecologist and recognise the existence of dynamic processes such as the three outlined above. In order to achieve preservation of a locality (or its predetermined modification) a process of *management* is needed and it is here that an ecological approach is of such importance.

An example will serve to illustrate the process in action. Chalk and limestone downland and the diverse flora and fauna associated with it, depends for its existence on the browsing of rabbits and the grazing of farm animals, particularly sheep. One of the commonest trees to colonise downland is hawthorn, *Crataegus monogyna*, which is dispersed by birds, particularly the greenfinch, *Carduelis chloris*, which eats the berries in autumn and passes the seeds out with the faeces. Provided grazing pressure remains high, few of the young seedlings are able to survive (Fig. 14.11(a)) but once it is relaxed the plant quickly spreads to form dense thickets (Fig. 14.11(b)). Management of downland therefore includes the systematic removal of unwanted hawthorn either by treatment with herbicides or, better, pulling up the plants by hand. Once a thicket has been allowed to develop, removal can be a difficult operation and the restoration of the former downland can take twenty years or more.

Some reasons for nature conservation

Conservation almost always involves a conflict of interests and its implementation usually demands a consideration not only of ecology, but also of influences such as politics, economics, and amenity. Suppose an area of land is designated an important nature reserve but is also part of a projected air field. Which claim should take precedence? Problems of this kind are not easy to resolve and it is important that both parties state their case as clearly as possible. What, then, is the case for nature conservation?

1 Preservation of rare species

The rate of disappearance of rare species of plants in Britain has been summarised by Perring [14.19]. During the period 1600–1900 only seven species are known to have become extinct, but from 1900–1970 a further thirteen disappeared.

(a)

(b)

Fig. 14.11 Stages in the colonisation of downland by hawthorn, *Crataegus monogyna*, (a) low density, (b) thicket formation

Moreover, an increasing number are falling into the category of 'very rare', that is to say restricted to one or two ten kilometre squares of the country. While in 1930 there were 59 in this group, forty years later there were 97. A similar situation exists for animals, particularly insects.

Putting extinction in perspective, we must bear in mind that throughout evolutionary time many more species have failed than have survived. Extinction is one outcome of the process of competition which has characterised plant and animal populations since their beginning. However, as we have seen, during the last few hundred years, under the adverse influence of man, the process of elimination has been greatly accelerated. Few would deny that threatened species such as the cheddar pink, *Dianthus gratianopolitanus*, and the large blue butterfly, *Maculinea arion*, are worthy of preservation not only for their intrinsic beauty, but also for the great biological interest associated with their life cycles and distribution. But whether this argument applies equally to every species on the verge of extinction is a matter for debate.

2 Preservation of communities

Just as individual species possess unique attributes of ecological interest, so, too, do communities of plants and animals. One of the richest in Britain is that of downland, which, as we have seen, depends upon grazing for its continued existence. In the absence of proper management, hawthorn scrub can quickly take over, a process which is occurring at present in many parts of the country. Another community at present under severe pressure is heathland (see Chapter 12). Spellerberg [14.20] has drawn attention to the increasing elimination and fragmentation of heaths in Dorset (Fig. 14.12) over the last 160 years due to the inroads of agriculture, afforestation, and urbanization. Thus, during the last 80 years, the area of uninterrupted heathland remaining dropped by about 80 per cent, while fragmentation over the same period increased by more than 300 per cent.

3 Provision of amenities

In recent years, great strides have been made to meet an increasing public demand for access to the countryside. Thus Duffey [14.10] records that today we have some 135 National Nature Reserves, 700 Naturalists' Trust Reserves, 50 reserves of the Royal Society for the Protection of Birds, 10 National Parks, and 166 660 ha (400 000 acres) of land owned by the National Trust, to mention only part of the range of amenities now open to public access. From this diversity of conservation experience one of the conclusions that has emerged is that public accessibility and the preservation of the aesthetic and ecological aspects of the countryside need not be in conflict *provided* suitable systems of management are adopted, such as the provision of parking facilities for cars and vigilance by wardens where rare species and communities are involved.

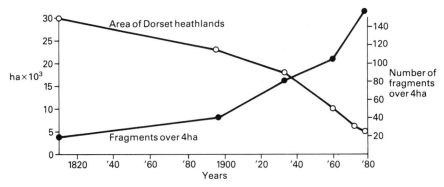

Fig. 14.12 Elimination and fragmentation of heathland in Dorset, 1820–1980 (after Spellerberg)

4 Areas for research and study

Since it is concerned with organisms in their natural environment, ecology is (or should be) a branch of science studied mainly out of doors. In this respect, nature reserves can provide outdoor laboratories. To meet this need the Nature Conservancy has established 3165 Sites of Special Scientific Interest (SSSI) throughout Britain. At one time these were regarded as places where, in the interest of ecological research, public access was sometimes restricted. Today, a more liberal view prevails and the management of many of them is comparable with that of other nature reserves.

5 Preservation of potentially useful species

The argument here is different from *1* above, in that many of the species likely to be of interest are quite common, although frequently with a restricted distribution.

(a) Species of genetic interest

A feature of agriculture and horticulture over the last 50 years or so has been the production of a host of new plant varieties (cultivars). Gardeners can now select their roses from a range of types, most of which were unknown at the beginning of the century. These differ markedly from their ancestors, some of which are no longer obtainable. These latter represent a collection of genotypes which may well be lost for ever, and it is now realised, a little belatedly, that steps should be taken to preserve such vanishing gene pools in case they are needed later. The same argument applies to wild species, a typical example being the beautiful little shrub, *Daphne mezereum*, with its pink, scented flowers and red berries. This is almost certainly an indigenous species, found in a few limestone localities, where it is no doubt reinforced by birds carrying seeds from gardens in which it is widely grown. Many of the garden forms are similar to the wild ones and are probably derived from them. However, modern selective breeding has produced cultivars which are more robust than their ancestors, having larger flowers with a different range of colours. These will no doubt eventually replace the wild form in cultivation. It is important that the wild form should be preserved, not only as an attractive wild species but also because the scent of its flowers is stronger than that of the cultivated varieties.

(b) Plants with potential food value

Some of the wild foods of Britain, such as blackberries and mushrooms, have traditionally formed part of the human diet. Others, such as nettle leaves, are more rarely eaten. With the increasing realisation that an excessive consumption of animal protein is not only economically prohibitive but can be detrimental to health, searches are being made for alternative naturally occurring foods. An example is the recent discovery that all parts of the common streamside plant comfrey, *Symphytum officinale*, contain unusual quantities of protein and therefore provide a potential future source. Findings such as this are unpredictable as they are related to the fluctuating pressures of human demand. It is therefore important that our nature reserves should be sufficiently diverse to include, as far as possible, the whole range of the British flora and fauna.

(c)   Plants of potential medicinal value
Traditionally, plant products have provided the basis of all medicine (the first synthetic drug was used in 1911). Classic examples are digitalin obtained from the foxglove, *Digitalis purpurea*, hyoscine from henbane, *Hyoscyamus niger*, and atropine from deadly nightshade, *Atropa bella-donna*. In all these the plant extract preceded the synthetic version as a curative agent. Today, interest in herbal medicine is still strong, and the number of plant-derived drugs is immense, as is illustrated by the fact that a reasonably complete modern herbal [14.21] runs to over 900 pages. The range of future possibilities may well be considerable, and this serves to reinforce the point made in the previous section that as great a diversity of plant habitats as possible must be preserved.

6 Economic aspects and tourism
Finally, there is the by no means unimportant fact that since our National Parks and other protected places are often areas of outstanding natural beauty, they provide a strong attraction to tourists both from Britain and abroad. Tourism is now an important national industry. Although the financial return to those most closely concerned with the maintenance of our reserves may be small, the overall effect on organisations such as hotels, catering, and transport can be immense. As we have seen, the interests of public access and conservation need not conflict, and the arrival of 1000 caravans in the vicinity of the New Forest over a summer's weekend is not catastrophic if management is effective, which it usually is.

Summary

1   It is doubtful if any ecological environment exists today that has not been affected somehow by the activities of man.

2   Pollutants of the atmosphere include numerous gases such as sulphur dioxide. Plants and animals have responded to such substances in a characteristic manner and some species such as lichens can be used as indicators of varying levels of pollution.

3   Some of the most extreme areas of pollution have occurred in our great rivers. The recovery of the River Thames to a level where it can be used for salmon fishing provides one of the success stories in conservation.

4   Some of the most successful pesticides developed for use in agriculture have also proved to be serious pollutants when used indiscriminately.

5   One of the most powerful influences of man is on the nature of ecological communities. The destruction of hedgerows and the establishment of conifer plantations are two examples.

6   The rate of consumption of essential resources by man has highlighted the need for recycling and the use of alternative materials that are not in such short supply.

7   The world-wide demand for protein is still increasing and in many developing countries there is widespread malnutrition. There is an urgent need for a more equitable distribution of available stocks and for improved methods of production. Changes in national feeding habits could play a significant part in alleviating shortage.

9   An ecological approach has been important in the control of disease, as in the reduction of malaria in Sri Lanka.

10 Ecology can be important in agriculture, for instance, in combating fungal disease in cereals, which is a major cause of loss.

11 Conservation of the environment is an active process necessitating management. Before instituting conservation measures it is important to be clear about the reasons behind them.

12 The economic and social aspects of amenity and tourism are of increasing importance. They have important ecological implications.

References

14.1 Richardson, D. H. S. (1975) *The Vanishing Lichens*, David and Charles

14.2 Laundon, J. R. (1967) 'A study of the lichen flora of London', *Lichenologist*, **3**, 277–327

14.3 Dalby, D. (1981) *Chart of Lichens and Air Pollution*, British Museum (Natural History)

14.4 Dowdeswell, W. H. (4th edn 1975) *The Mechanism of Evolution*, Heinemann

14.5 Kettlewell, H. B. D. (1958) 'A survey of the frequencies of *Biston betularia* and its melanic forms in Great Britain', *Heredity*, **12**, 51–72

14.6 Kettlewell, H. B. D. (1956) 'Further selection experiments on industrial melanism in the Lepidoptera', *Heredity*, **10**, 287–301

14.7 Ford, E. B. (4th edn 1975) *Ecological Genetics*, Chapman & Hall

14.8 Doxat, J. (1977) *The Living Thames*, Hutchinson Benham

14.9 Yapp, W. B. (1972) *Production, Pollution, Protection*, Wykeham

14.10 Duffey, E. (1974) *Nature Reserves and Wild Life*, Heinemann

14.11 Edlin, H. L. (1956) *Trees, Woods and Man*, Collins

14.12 Elton, C. S. (1958) *The Ecology of Invasions by Animals and Plants*, Methuen

14.13 Cloud, P. (1969) *Resources and Man*, Freeman

14.14 Taylor, T. G. (1982) *Nutrition and Health*, Arnold

14.15 Bowman, J. C. (1977) *Animals for Man*, Arnold

14.16 Dowdeswell, W. H. (1981) *The Protein Project, An Example of Commercial Biosynthesis*, B.P. Educational Service

14.17 Fletcher, W. W. (1974) *The Pest War*, Blackwell

14.18 Perring, F. H. and Mellanby, K. (Eds) (1977) *Ecological Effects of Pesticides*, Academic Press

14.19 Perring, F. H. (1970) 'The last seventy years', in *The Flora of a Changing Britain*, Botanical Society of the British Isles Report No. 11

14.20 Spellerberg, I. F. (1981) *Ecological Evaluation for Conservation*, Arnold

14.21 Grieve, M. (1976) *A Modern Herbal*, Penguin

Bibliography

The titles included here are some of those widely used in the study of ecology. The list makes no claim to completeness and in some areas of the subject such as the identification of birds and flowering plants, suitable alternatives are available at all levels. References on particular aspects of the subject are also included at the end of each chapter.

General ecology

Bennett, D. P. and Humphries, D. A. (2nd edn 1983) *Introduction to Field Biology*, Arnold
Colinvaux, P. A. (1973) *Introduction to Ecology*, Wiley
Darlington, A. and Leadley Brown, A. (1975) *One Approach to Ecology*, Longmans
Elton, C. (1958) *The Ecology of Invasions by Animals and Plants*, Methuen
King, T. J. (1980) *Ecology*, Nelson
Odum, E. P. (3rd edn 1971) *Fundamentals of Ecology*, Saunders
Owen, D. F. (1974) *What is Ecology?*, Oxford
Ricklefs, R. E. (1973) *Ecology*, Nelson
Sands, M. K. (1978) *Problems in Ecology*, Mills and Boon

Animal ecology

Colinvaux, P. A. (1980) *Why Big, Fierce Animals are Rare*, Penguin
Elton, C. S. (1927) *Animal Ecology*, Sidgwick and Jackson
Elton, C. S. (1966) *The Pattern of Animal Communities*, Methuen
MacFadyen, A. (2nd edn 1963) *Animal Ecology, Aims and Methods*, Pitman
Varley, G. C., Gradwell, G. R. and Hassell, M. P. (1980) *Insect Population Ecology*, Blackwell

Plant ecology

Ashby, M. (2nd edn 1969) *Introduction to Plant Ecology*, Macmillan
Billings, W. D. (2nd edn 1972) *Plants, Man and the Ecosystem*, Macmillan
Bishop, O. N. (1973) *Natural Communities*, John Murray
Tivy, J. (2nd edn 1982) *Biogeography, A study of Plants in the Ecosphere*, Longmans

Soil ecology

Jackson, R. N. and Raw, F. (1966) *Life in the Soil*, Studies in Biology No. 2, Arnold
Leadley Brown, A. (1978) *Ecology of Soil Organisms*, Heinemann
Russell, E. J. (1957) *The World of Soil*, New Naturalist, Collins

Russell, E. J. (1961) *Soil Conditions and Plant Growth*, Longmans
Stamp, L. D. (1962) *The Land of Britain: Its Use and Misuse*, Longmans
Townsend, W. N. (1973) *An Introduction to the Scientific Study of Soil*, Arnold
Wallwork, J. A. (1970) *Ecology of Soil Animals*, McGraw-Hill

Freshwater ecology

Clegg, J. (4th edn 1974) *Freshwater Life*, Warne
Clegg, J. (3rd edn 1980) *The Observers Book of Pond Life*, Warne
Hynes, H. B. N. (1971) *The Ecology of Running Waters*, University of Liverpool Press
Leadley Brown, A. (1971) *Ecology of Fresh Water*, Heinemann
Macan, T. T. (1963) *Freshwater Ecology*, Longmans
Mills, D. N. (1972) *An Introduction to Freshwater Ecology*, Oliver & Boyd
Townsend, C. R. (1980) *The Ecology of Streams and Rivers*, Studies in Biology No. 122, Arnold

Marine and estuarine ecology

Barnes, R. S. K. (1974) *Estuarine Biology*, Studies in Biology No. 49, Arnold
Barrett, J. H. and Yonge, C. M. (1964) *Pocket Guide to the Sea Shore*, Collins
Bradfield, A. E. (1978) *Life on Sandy Shores*, Studies in Biology No. 89, Arnold
Brehant, R. N. (1982) *Ecology of Rocky Shores*, Studies in Biology No. 139, Arnold
Campbell, A. C. and Nicholls, J. (1976) *Hamlyn Guide to the Seashore* and *Shallow Seas of Britain and Europe*, Hamlyn
Jenkins, M. (1983) *Seashore Studies*, George Allen & Unwin
McLusky, D. S. (1971) *Ecology of Estuaries*, Heinemann
Southwood, A. J. (1965) *Life on the Sea Shore*, Heinemann

Ecological investigation

Bailey, N. T. J. (1959) *Statistical Methods in Biology*, English Universities Press
Begon, M. (1979) *Investigating Animal Abundance*, Arnold
Bishop, O. N. (3rd edn 1981) *Statistics for Biology*, Longmans
Campbell, R. C. (2nd edn 1974) *Statistical Methods in Biology*, Cambridge
Finney, D. J. (1980) *Statistics for Biologists*, Chapman and Hall
Grieg-Smith, P. (2nd edn 1964) *Quantitative Plant Ecology*, Butterworth
Heath, O. V. S. (1970) *Investigation by Experiment*, Studies in Biology No. 23, Arnold
Lewis, T. and Taylor, L. R. (1967) *Introduction to Experimental Ecology*, Academic Press
Mills, D. H. (1972) *An Introduction to Freshwater Ecology*, Oliver & Boyd
Parker, R. E. (2nd edn 1980) *Introductory Statistics for Biology*, Studies in Biology No. 43, Arnold
Southwood, T. R. E. (2nd edn 1978) *Ecological Methods*, Chapman & Hall
Wratten, S. D. and Fry, G. L. A. (1980) *Field and Laboratory Exercises in Ecology*, Arnold

Identification of land plants

Chancellor, R. J. (1966) *The Identification of Weed Seedlings of Farm and Garden,* Blackwell

Darlington, A. and Hirons, M. J. D. (1968) *The Pocket Encyclopaedia of Plant Galls,* Blandford

Fitter, R., Fitter, A. and Blamey, M. (3rd edn 1978) *The Wild Flowers of Britain and Northern Europe,* Collins

Lange, M. and Hora, F. B. (2nd edn 1981) *Mushrooms and Toadstools,* Collins

Martin, W. K. (3rd edn 1978) *The Concise British Flora in Colour,* Sphere

Mitchell, A. (1974) *A Field Guide to the Trees of Britain and Northern Europe,* Collins

Nicholson, B. E. and Brightman, F. H. (1979) *The Oxford Book of Flowerless Plants,* Oxford

Phillips, R. (1980) *Grasses, Ferns, Mosses and Lichens of Britain and Ireland,* Pan/Ward Lock

Phillips, R. (1978) *Trees in Britain, Europe and North America,* Pan/Ward Lock

Phillips, R. (1977) *Wild Flowers in Britain,* Pan/Ward Lock

Polunin, O. (1977) *Trees and Bushes of Britian and Europe,* Paladin

Rayner, R. (1979) *Mushrooms and Toadstools,* Hamlyn

Identification of freshwater plants and animals

Clegg, J. (3rd edn 1980) *The Observer's Book of Pond Life,* Warne

Leadley Brown, A. M. (1970) *Key to Pond Organisms,* Nuffield Advanced Biological Science, Penguin

Macan, T. T. (1959) *A Guide to Freshwater Invertebrate Animals,* Longmans

Mellanby, H. (6th edn 1963) *Animal Life in Fresh Water,* Methuen

Quigley, M. (1977) *Invertebrates of Streams and Rivers, A Key to Identification,* Arnold

For keys to specific groups of aquatic invertebrates, see the publications of the Freshwater Biological Association, The Ferry House, Ambleside, Cumbria, LA22 0LP.

Identification of sea shore plants and animals

Barrett, J. H. and Yonge, C. M. (1958) *Pocket Guide to the Sea Shore,* Collins

Evans, S. M. and Hardy, J. M. (1970) *Sea Shore and Sand Dunes,* Heinemann

Overend, D. and Barrett, J. (1981) *Collins Handguide to the Sea Coast,* Collins

Identification of land animals

Vertebrates

Bang, P. and Dahlstrom, P. (1980) *Animal Tracks and Signs,* Collins

Mammals

Corbet, G. and Ovenden, D. (1980) *The Mammals of Britain and Europe,* Collins

Southern, H. N. (Ed) (1964) *The Handbook of the British Mammals,* Blackwell

Birds
Gooders, J. (1968) *Where to Watch Birds*, Andre Deutsch
Hammond, N. and Everett, M. (1980) *Birds of Britain and Europe*, Pan/Ward Lock
Heinzel, H., Fitter, R. and Parslow, J. (3rd edn 1977) *The Birds of Britain and Europe*, Collins
Peterson, R., Mountford, G., and Hollom, P. A. D. (3rd edn 1979) *A Field Guide to the Birds of Britain and Europe*, Collins

Amphibians and Reptiles
Arnold, E. N. and Burton, J. A. (1978) *A Field Guide to the Reptiles and Amphibians of Britain and Europe*, Collins
Smith, M. (1951) *The British Amphibians and Reptiles*, New Naturalist, Collins

Fishes
Vevers, G. (2nd edn 1978) *Fishes in Colour: Marine and Freshwater*, Witherby

Invertebrates

Cloudsley-Thompson, J. L. and Sankey, J. (1961) *Land Invertebrates*, Methuen
Nichols, D. and Cooke, J. (1979) *The Oxford Book of Invertebrates*, Oxford
Paviour-Smith, K. and Whittaker, J. B. (1969) *A Key to the Major Groups of British Free-living Terrestrial Invertebrates*, Blackwell

Annelids
Edwards, C. A. and Lofty, J. R. (1967) *Biology of Earthworms*, Chapman and Hall

Arthropods
(i) Arachnids
Savory, T. H. (2nd edn 1945) *The Spiders and Allied Orders of the British Isles*, Warne

(ii) Crustaceans
Edney, E. B. (1954) *British Woodlice*, Synopses of the British Fauna No. 9, Linnean Society, London

(iii) Insects
Burton, J. (1981) *The Oxford Book of Insects*, Oxford
Chinery, M. (2nd edn 1979) *A Field Guide to the Insects of Britain and Northern Europe*, Collins
Chu, H. F. (1949) *How to Know the Immature Insects*, Brown, Iowa
Colyer, C. N. and Hammond, C. O. (1951) *Flies of the British Isles*, Warne
Corbet, P. S., Longfield, C. and Moore, N. W. (1960) *Dragonflies*, New Naturalist, Collins
Friedlander, C. P. and Priest, D. A. (1955) *Insects and Spiders*, Pitman
Higgins, L. G. and Riley, N. D. (1970) *A Field Guide to the Butterflies of Britain and Europe*, Collins
Linssen, E. F. (1959) *Beetles of the British Isles*, Warne
Longfield, C. (1937) *Dragonflies of the British Isles*, Warne
South, R. (3rd edn 1943) *The Moths of the British Isles*, (2 volumes) Warne

Southwood, T. R. E. (1959) *Land and Water Bugs of the British Isles*, Warne

Step, E. (1932) *Bees, Wasps, Ants and Allied Orders of the British Isles*, Warne

Stokoe, W. J. (1948) *The Caterpillars of the British Moths* (2 volumes), Warne

Stokoe, W. J. (1948) *The Caterpillars of the British Butterflies*, Warne

Stokoe, W. J. (1979) *The Observer's Book of Butterflies*, Warne

For keys to the insect orders, see the publications of the Royal Entomological Society, 41 Queen's Gate, London SW7 5HU

(iv) Myriapods

Blower, J. G. (1958) *British Millipedes (Diplopoda)*, Synopses of the British Fauna No. 11, Linnean Society, London

Cloudsley-Thompson, J. L. (1958) *Spiders, Scorpions, Centipedes and Mites*, Pergamon

Eason, E. H. (1964) *Centipedes of the British Isles*, Warne

Molluscs

Janus, N. (1965) *Land and Freshwater Molluscs*, Burke

Kerney, M. P., Cameron, R. A. D. and Riley, G. (1979) *A Field Guide to the Land Snails of Britain and Northwest Europe*, Collins

McMillan, N. F. (1968) *British Shells*, Warne

Nematodes

Goodey, T. (1951) *Soil and Freshwater Nematodes*, Methuen

Some useful addresses

Note: The following list owes much to a similar one included in the Teachers' Guide of the publication: Barker, J. A. *et al.* (1975) *People and Resources*, Ed Kelly, Evans Brothers Ltd, London. The four books (three texts and a teacher's guide) contain a wealth of information and ideas on man and the environment. They are strongly recommended as a source for further reference.

Association of Agriculture Victoria Chambers, 16–20 Strutton Ground, London SW1P 2HP (Advice on agricultural topics)

Association for Science Education College Lane, Hatfield, Herts, AL10 9AA (Publisher *School Science Review* and *Education in Science*)

British Ecological Society Department of Biology, The University, Heslington, York, YO1 5DD

British Petroleum Co. Ltd B.P. Educational Service, Britannic House, Moor Lane, London EC2Y 9BU (Publications on the environment)

British Waterworks Association Information Department, 4 Park Street, London W1Y 4BL (Publications on water conservation)

Conservation Society 34 Bridge Street, Walton-on-Thames, Surrey, KT12 1AJ (Publications on resources)

Council for Nature c/o Zoological Society of London, Regent's Park, London NW1 4RY (Coordinates information on conservation)

Council for Environmental Education School of Education, University of Reading, 24 London Road, Reading, Berkshire, RG1 5AQ (Publishes *Directory of Environmental Literature* and *Ecological Teaching Aids*)

Countryside Commission John Dower House, Crescent Place, Cheltenham, Gloucestershire GL50 8RA

Esso Petroleum Co. Ltd Education Service, Esso House, Victoria Street, London SW1E 5JW (Publications on the environment)

Fauna Preservation Society c/o Zoological Society of London, Regent's Park, London NW1 4RY (Publishes journal *Oryx*, concerned with preservation of world fauna)

Field Studies Council 9 Devereux Court, Strand, London WC2R 3JR (Publishes journal and operates numerous field centres)

Fisons Fertiliser Division Harvest House, Cobbold Road, Felixstowe, Suffolk, IP11 7LP (Information on fertilizers and the environment)

Freshwater Biological Association Windermere Laboratory, The Ferry House, Far Sawrey, Ambleside, Cumbria, LA22 0LP

Friends of the Earth 9 Poland Street, London W1V 3DS (Action group, publications on the environment)

Imperial Chemical Industries Agricultural Division, Publicity Department, PO Box 1, Billingham, Teesside, TS23 1LB (Information on fertilisers and the environment)

Institute of Biology 20 Queensberry Place, London SW7 2DZ (Publishes *Biologist* and *Journal of Biological Education*, includes education and environment divisions)

Marine Biological Association The Laboratory, Citadel Hill, Plymouth, Devon, PL1 2PB

Thames Water New River Head, Rosebery Avenue, London EC1R 4TP (Publications on the conservation of the River Thames)

Murphy Chemical Co. Ltd Wheathampsted, St. Alban's, Herts, AL4 8QU (Information on pesticides)

National Association for Environmental Education Information Officer, 18 Barrowdale Close, Exmouth, Devon EX8 5PN (Environmental education at all levels)

Nature Conservancy 19 Belgrave Square, London SW1X 8PY (Information on nature reserves and conservation research)

Royal Entomological Society 41 Queen's Gate, London, SW7 5HU (Publishes useful keys to most groups of insects)

Royal Society for the Protection of Birds The Lodge, Sandy, Bedfordshire, SG19 2DL (Wide range of publications on birds, their habitats, and distribution)

School Natural Science Society Publications Officer, 44 Claremont Gardens, Upminster, RM14 1DN (Publishes *Natural Science in Schools* and a wide range of leaflets, some ecological)

Shell International Petroleum Co. Ltd Shell Education Service, Shell Centre, London SE1 7NA (Publishes *Aids for Teachers* and information on the environment)

Society for the Promotion of Nature Reserves, and **Association of Nature Conservation Trusts** The Manor House, Alford, Lincs, LN13 9DL (Information on all nature reserves particularly those of County Naturalist Trusts)

Soil Association Walnut Tree Manor, Haughley, Stowmarket, Suffolk, IP14 3RS (Numerous publications on soil, interest in organic culture)

Town and Country Planning Association 17 Carlton House Terrace, London SW1Y 5AS (Publishes *Bulletin of Environmental Education*)

Wildlife Youth Service Marston Court, Manor Road, Wallington, Surrey, SM6 0DN (A junior branch of the World Wildlife Fund)

World Wildlife Fund Panda House, 11–13 Ockford Road, Godalming, Surrey GU7 1QU (Concerned with the preservation of wildlife throughout the world)

Glossary

Abiotic factors Non-biological factors such as temperature, that form part of the environment of an organism (synonymous with **physical factors**)

Adaptability Capacity for evolutionary change. This may depend on the reaction of the phenotype to environmental variation as well as to the overall genetic variability of the population

Adaptation A genetically determined characteristic that governs the capacity of an organism to survive in its environment

Allopatric population A population which is divided by geographic boundaries (see **sympatric population**)

Analysis of variance A statistical procedure for comparing more than two means

Autecology The ecological relationships of a particular plant or animal species (synonym for **population ecology**)

Autotrophic Producing food using the light energy of the sun through photosynthesis, as in green plants and some bacteria (see **heterotrophic**)

Barrier An ecological factor which restricts the range of a species

Biological control The use of natural predators or parasites to control pests

Biomass The mass of living material usually as dry weight. Commonly expressed as weight per unit area

Biosphere That part of the earth and its environment that is inhabited by living organisms

Biotic factors Biological factors such as food and living space which result from the interaction of living organisms with one another

Calcicole A plant that thrives on neutral or alkaline soils (see **calcifuge**)

Calcifuge A plant that thrives on lime-free acidic soils (see **calcicole**)

Carbon cycle The circulation of carbon between living organisms and the environment where it occurs in the form of the gas carbon dioxide

Chi-squared (χ^2) A statistic used to measure the probability of one ratio deviating from another in accordance with some hypothesis

Climax The last stage of a sere and the end of a successional sequence under a particular set of environmental conditions

Cline A gradient of changes in the characteristics of a population taking place over a geographical area

Clumped distribution A numerical arrangement in which the different values are not spaced out uniformly

Commensalism A situation in which two organisms (**commensals**) belonging to different species live together without becoming physiologically interdependent

Community A group of interacting populations. Its range is usually determined by the nature of their interactions and their spatial occurrence (see **population**)

301

Compensation period   The time taken to reach the compensation point (see below)

Compensation point   The light intensity at which the rate of photosynthesis is balanced by that of respiration so that the net exchange of oxygen and carbon dioxide is zero.

Competition – interspecific   The struggle for survival taking place between individuals of different species

Competition – intraspecific   The struggle for survival taking place between individuals of the same species

Conservation   The artificial control of an ecological environment in a particular state of balance between the various species present

Correlation coefficient   A statistic used to measure the degree of relationship existing between two variables

Cryptic colouration   Colours that tend to match the environment in which an animal lives (c.f. **aposematic** colours that advertise an animal's presence)

Cyanogenic plant   A plant that produces hydrogen cyanide. This can usually be detected with sodium picrate paper when the plant is cut or bruised

Datum level   The level used as a reference in surveying

Degrees of freedom   The least number of independent variables that must be given values before the state of a quantitative system can be completely determined

Density   The number of individuals per unit area

Density dependent factors   Factors influencing a population whose effects vary with the size of the population, e.g. living space

Density independent factors   Factors influencing a population whose effects do not vary with the size of the population, e.g. temperature

Detritivores   Animals whose principal food is detritus (see below)

Detritus   Semi-decomposed organic matter. A term most commonly used in describing aquatic habitats (equivalent to **humus** in soil)

Dispersion   The variation of a series of values

Diversity index   A parameter used to quantify the diversity of a community

Ecological genetics   An approach to ecology that takes account of the genetic and evolutionary aspects

Ecological niche   The economic status occupied by an organism within a community resulting from its adaptation to a particular set of environmental conditions, particularly food supply

Ecology   The relationships of living organisms with one another and with their environment

Ecosystem   A unit representing the interaction of all the living and non-living components in a particular locality

Ectoparasite   A parasite living on the exterior of its host (see **endoparasite**)

Edaphic factors   Factors relating to conditions in the soil

Electrophoresis   The movement of electrically charged particles towards oppositely charged electrodes in a solution subjected to an electric field. The rate of migration varies with molecule size

Endoparasite   A parasite living inside the body of its host (see **ectoparasite**)

Epibiont   Any organism that lives on the outside of another irrespective of its relationship with that organism

Epiphyte A plant that lives wholly, but not parasitically, on other plants

Epizoite An animal that lives wholly, but not parasitically, on other animals. Epizoites are usually able to lead an independent existence

Euryhaline Able to tolerate a wide range of external osmotic changes (see **stenohaline**)

Eurythermous Able to tolerate a wide range of temperature variation (see **stenothermous**)

Eutrophic Description of an area of water containing abundant nutrients and with a high productivity which cause lack of oxygen in lower levels

Extinction point The lowest percentage of full daylight at which a plant species is able to survive under natural conditions

Facultative parasite A parasite which is also capable of leading a free-living existence

Food chain A representation of the passage of food and other resources through populations within a community

Food web A representation of the relationships between a number of different food chains

Genetic drift Variations in gene frequency in small populations due to their size

Habitat The place where an organism lives

Halophyte A plant capable of tolerating soils containing a high concentration of salt, e.g. salt marshes

Heterotrophic Organisms which use complex organic materials for food, as in herbivores and carnivores (see **autotrophic**)

Holism An approach which takes into account the situation as a whole (see **reductionism**)

Holocoenotic relationships Relationships between living organisms regarded as a continuum with no barriers between them. (A term coined by Karl Friedrich in 1927)

Homeostasis Maintenance of constant internal conditions in the face of a fluctuating external environment

Host A living organism at whose expense a parasite exists or on whom an epibiont lives

Humus Semi-decomposed organic matter in soil

Hydrarch succession Succession beginning with an aquatic community (synonymous with **hydrosere**)

Hydrological cycle The circulation of water in nature including its interchange between earth and sky

Hydrosere Succession beginning with an aquatic community (see **xerosere**)

Hydrotaxis The movement of an animal towards (positive) or away from (negative) a source of water

Hyperparasite An organism parasitising another which is itself a parasite

Hypertonic Having a higher osmotic pressure

Hypotonic Having a lower osmotic pressure

Indicator species Species of plants which characterise a particular kind of habitat

Indigenous species   Species regarded as natives of a particular area
Integrated control   The control of pests by a combination of chemical and biological methods used together
Interspecific   Relating to interactions taking place between individuals of different species
Intraspecific   Referring to interactions occurring between individuals of the same species
Isotonic   Having the same osmotic pressure

Leaching   Removal of soluble compounds from soil by water, mainly rain
Litter layer   A layer of organic material on the surface of soil
Littorine species   Species inhabiting the sea shore

Macroclimate   The climate of a major habitat
Mean   The average of a series of numbers
Mechanical analysis   The process of separating and measuring the solid constituents of soil
Melanism   The possession by an animal of the nitrogenous pigment melanin giving it a brown or black appearance
Microclimate   The climate of a microhabitat
Microhabitat   A particular part of a habitat where an individual species (usually an animal) is normally found
Mor   Acid soil in which the process of organic decay has been greatly slowed down
Mull   A moist, well-aerated soil favouring the rapid breakdown of leaf litter by bacteria to form humus
Mutualism   A relationship between two or more species that is a benefit to all parties
Mycorrhiza   The close association between fungi and the roots of trees and shrubs which facilitates the uptake of minerals

Neap tides   Tides that occur between successive spring tides (see **spring tides**)
Nitrification   The conversion of ammonia to nitrite, and nitrite to nitrate by bacteria in soil and water.
Nitrogen cycle   The circulation of nitrogen between living organisms and the environment
Nitrogen fixation   The formation of nitrogenous compounds from atmospheric nitrogen
Normal distribution   A distribution of values which conforms to a bell-shaped curve
Null hypothesis   A hypothesis that no relationship exists between two variables (e.g. in calculations using chi-squared)

Osmoregulation   The process of regulating osmotic pressure

Pan   Leached material which has accumulated below the soil surface as a hard impervious layer
Parasite   An organism that lives either partially or completely at the expense of another living animal or plant (the **host**)

Parasite chain The food relationships existing between different parasites and their hosts

Photosynthesis A process occurring in green plants whereby sugar is formed in the presence of sunlight, a by-product being oxygen

Phototaxis The movement of an animal towards (positive) or away from (negative) a source of light

Phototropism The bending of a static organism towards (positive) or away from (negative) a source of light

Physical factors Non-biological factors such as temperature that form part of the environment of an organism (synonymous with **abiotic factors**)

Plankton Minute drifting aquatic plants (phytoplankton) and animals (zooplankton)

Podsol Cold, poorly aerated, and sometimes waterlogged soil in which the rate of bacterial decomposition of organic matter is much reduced

Polygenic inheritance A genetic situation in which a single character is controlled by two or more pairs of alleles. The result is a continuously graded type of variation

Polymorphism The occurrence together of two or more forms of the same species

Polytypic species A species in which there is a considerable degree of variation

Population A group of organisms of the same species occupying a particular ecological area (see **community**)

Primary consumers Herbivorous animals

Primary producers Organisms which produce their food from simple organic molecules by photosynthesis or chemosynthesis (**autotrophs**)

Probability The chance that a particular event will occur

Production – gross In green plants this is the total amount of light fixed through photosynthesis in a given period of time

Production – net Gross production less the amount of energy needed for respiration. The remainder is theoretically available for growth

Productivity The level of production of organic material in green plants which leads to different rates of growth

Propagule An agent of propagation such as the seed of a flowering plant

Protocooperation An association between two species which is beneficial to both but not obligatory

Psammosere The sequence of succession that takes place on sand dunes

Pyramid of biomass A figurative method of representing the relationship between trophic levels and the mass of organisms (dry weight) in an ecosystem

Pyramid of numbers A figurative method of representing the relationship between trophic levels and number of organisms in an ecosystem

Quadrat A square of known size. Used to determine the distribution and density of plants in different localities

Reductionism An approach which involves reducing a situation to its component parts (see **holism**)

Respiration A metabolic process involving the breakdown of food, with or without oxygen, and the release of energy

Rheotaxis A response by an animal to the impact of a water current. Positive rheotaxy denotes a tendency to move upstream, negative rheotaxy, downstream

Secondary consumers Carnivores that prey upon herbivores

Secondary host An organism other than a parasite's primary host in which the life cycle of the parasite is continued. The secondary host often acts as a means of transmission and dispersal (see **vector**)

Sere A sequence of plant succession

Significance In statistics, a measure of the reliability of a difference between observation and expectation, expressed as a probability

Specialisation Structural or physiological characteristics in an organism which enable it to survive in a peculiar set of environmental conditions

Specificity The dependence of an organism for its existence on a close physiological relationship with another species, e.g. parasite and host

Spring tides The most extreme tides occurring fortnightly at the time of the new and full moon (see **neap tides**)

Standard deviation A statistical measure of dispersion about a mean. It is represented by the square root of the variance

Standard error A statistical quantity used to compare two variables and to estimate the probability that a given result will be equalled or surpassed by chance

Standing crop The mass of living material present in a community at a particular moment

Stenohaline Able to tolerate only a narrow range of external osmotic changes (see **euryhaline**)

Stenothermous Able to withstand only small variations in temperature (see **eurythermous**)

Stratification As an ecological term, the zonation of vegetation such as the different layers in woodland

Succession A progressive series of changes in the plant and animal life of a community from initial colonisation to climax

Symbiosis A situation in which two organisms (**symbionts**) exist together in close physiological union to their mutual benefit. The condition is usually obligatory

Sympatric population A population not divided by geographical barriers (see **allopatric**)

Synecology The ecological relationships of a community (synonymous with **community ecology**)

Synthesis The building up of a substance or situation, in contrast to **analysis** which involves a breakdown into smaller component parts

System Any entity consisting of a number of interacting parts. The removal or failure of one part may incapacitate the whole system

Systematics The process of classifying living organisms

Taxonomy The practical process of description, classification, and naming of plants and animals

Territory An area established and defended by a pair of animals usually during the breeding season. In birds it is the locality within which the nest is built

Territorial behaviour The characteristic pattern of behaviour associated with the establishment and maintenance of a territory. It is particularly well developed in birds

Tertiary consumers Carnivores preying upon other carnivores. They represent the top trophic level of a food web

Thermistor A semiconductor consisting of a complex metal oxide whose resistance decreases with increasing temperature. Used in electronic thermometers

Thermotaxis The movement of an animal towards (positive) or away from (negative) a source of heat

Thigmotaxis A response by an animal to the stimulus of touch

Transect A method of measuring and representing graphically the distribution of plants and animals. Belt transects measure horizontal distribution within a band (often one metre wide). Line transects are useful in obtaining a profile of vegetation

Trophic level The level in the food web at which a group of organisms occurs. Green plants (primary producers) are at the lowest level; tertiary consumers at the highest level

Uniform distribution The equal distribution of a species in the areas where it occurs

Variance A measure of the deviation from the mean, given by the mean of the squares of the deviations (s^2)

Vector An animal other than the host which acts as a parasite's means of transmission and dispersal, and in which part of the life cycle *does not* take place (see **secondary host**)

Xerophyte A plant possessing particular characteristics which enable it to withstand drought

Xerarch succession A succession in which the original habitat was dry (see **hydrarch succession**)

Xerosere Synonym for **xerarch succession** (see **hydrosere**)

Index

Abiotic factor, *see* physical factor
Acorn barnacle, 214
Activity, animal, 17–19
Adaptability, 11
Adaptation, 159–60, 202–5
Admiralty Tide Table, 121
Adonis blue butterfly, 168
Aeshna sp. (dragonfly), 245
African migratory locust, 30
Agricultural practice, 274–80, 284, 286–8
Agrostis spp. (grass), 24–6, 166, 239
Allee, W. C., 30
Allolobophora nocturna (earthworm), 81
Amensalism, 40
Analysis, 3
Annual meadow grass, 251
Anthoxanthum odoratum (grass), 24
Anvil stone, of thrush, 168, 169, 259
Arctic fox, 31
Armadillidium (woodlouse), 153
Ash, 168, 239, 250
Ashby, M., 87–8, 159
Association, of species, 20–1
Atmometer, 108
Autotroph, 50, 68

Badger, 51
Baermann funnel, 84–5, 265
Bailey's triple catch, estimation of numbers, 94–5
Baldwin, E., 229
Baltic tellin, 233
Bar chart, 128–9
Beating apparatus, 77
Behaviour pattern, 4, 46, 203–4
Bell heather, 21, 236, 247
Benomyl, 287
Benzene hexachloride (BHC), 286
Berrie, A. D., 58
Bignell beating tray, 77
Bilberry, 239, 247
Binomial system, 5
Biological control, of pests, 41–5
Biome, 49
Biometry, 2
Biosphere, 149
Biotechnology, 284–5
Birch, 239
Bird'sfoot trefoil, 25, 167, 171, 239, 240, 275
Birth control, 46
Bishop, O. N., 108, 254, 263
Blackthorn, 250

Bladder campion, 25
Bladderwort, 241, 243
Bluebell, 157, 252
Bog, 151, 237; asphodel, 245; blanket, 245; community, 243–6; moss, 236, 245; myrtle, 243
Bomb calorimeter, 57
Bowman, J. C., 283
Bracken, 239
Bradshaw, A. D., 24
Bramble, 156, 168
Bristle-pointed hair moss, 240
Broadback wheatfield, Rothamsted, 10
Brook lamprey, 202
Broom, 240
Brown-lipped snail, 251
Bryum spp. (moss), 255
Bullhead, 205
Bulrush, 191, 196
Bureau of Animal Population, Oxford, 12
Butterwort, 241, 243, 246

Calcicole plant, 112, 166, 181, 189, 251, 254
Calcifuge plant, 112, 239, 251, 254
Calanus sp. (crustacean), 58, 64, 189
Caloplaca (lichen), 269
Campion, 251
Canadian pondweed, 191, 192, 205
Capercailzie, 277
Carchesium sp. (protozoan), 54
Carp, 201
Carrion crow, 4
Carteria sp. (alga), 41
Catworm, 233
Cavity chamber, 93
Celandine, 157
Cereal, fungal disease, 286–8
Chaffinch, 278
Chalkhill blue butterfly, 169
Chameleon shrimp, 230
Cheddar pink, 290
Chemosynthetic organism, 61
Chi-squared (χ^2), 15, 138–42, 148
Chironomus sp. (midge), 115, 185, 188, 193, 245
Chloeon dipterum (dragonfly), 186
Chlorides, estimation, 118–19
Chlorohydra viridissima (coelenterate), 40
Cinquefoil, 263
Cinnabar moth, 10
Cladonia spp. (lichen), 240

Classification, 3
Clatworthy, J. N., 29
Clay, in soil, 103
Climax, 10, 22, 67, 68, 70
Cline, 5
Clover, 171, 259
Cobalt chloride, 109; thiocyanate, 109
Cockle, 214, 224, 231, 233
Coenagrion spp. (dragonfly), 245
Colgan, N., 221
Colinvaux, P., 54
Collecting apparatus, 72–85
Colonisation, process of, 7–9
Columella aspersa (snail), 245
Comfrey, 36, 291
Commensalism, 40
Common blue butterfly, 168, 275
Common duckweed, 29
Common polypody, 15
Common rush, 252
Commonwealth Institute of Biological Control, 45
Community, 149, 205
Compensation period, 159; point, 159
Competition, 40, 66–7, 144; interspecific, 8, 10, 32, 67, 70; intraspecific, 1, 8, 9, 38, 40, 66, 70
Computers, in ecology, 142–4
Conductivity, measurement, 119
Conifer, 276–8
Conservation, 2, 288
Consumer, primary, 9, 50–1, 283; secondary, 9, 50–1, 283; tertiary, 10, 50
Cook, L. M., 37
Cordgrass, 228, 229, 231, 232
Corixa sp. (bug), 187, 246
Corvus sp., *see* crows
Correlation, 136–8
Correlation coefficient, 137–8, 147
Cothill, Berkshire, 36
Cotton grass, 236, 243
Creeping buttercup, 259
Creeping thistle, 262
Crellin, J. R., 107, 117, 118
Crenobia alpina (flatworm), 201
Cross-leaved heath, 245, 247
Crow, 4
Crowson, R. A., 3
Cryptic colouration, 213
Cryptozoic species, 109
Cultivar, 291

Current, as an ecological factor, 109–12, 201–4; measurement, 109–12
Cyanogenic plant, 171
Cycle, carbon, 59; global, 58; hydrological, 58–9; localised, 61–2; nitrogen, 60–1; nutrients, 58–62; sedimentary, 61–2; sulphur, 61–2
Cyclops sp. (crustacean), 19, 69, 183

DDT, 9, 274, 285; resistance, 9, 275
Dactylis glomerata (grass), 25
Daisy, 260
Dalby, C., 271
Dandelion, 254, 259
Daphne mezereum (flowering plant), 291
Daphnia spp. (crustacean), 18, 69, 183, 185
Darlington, A., 243, 254, 255, 258, 263
Darwin, C., 149, 150
Datum level, 120
Davison, A. W., 233
Dawson, F. H., 108
Deadly nightshade, 292
Debenham level, 122
Decomposer, 61
Degrees of freedom, 135
Dendrocoelum lacteum (flatworm), 189
Density, control, 50, 87; determination, 143; of population, *see* population; variations, 31–8
Density-dependent factor, 8, 29–31
Density-independent factor, 31, 37
Department of Health and Social Security, 282
Detritus, 53, 60, 202
Detritivore, 60
Direct counts, of populations, 91–3
Desiccation, 214, 240
Dieldrin, 274, 286
Dispersion, arithmetic, 85, 87; clumped, 21; of species, 20–1; random, 20; uniform, 21
Distribution, animal, 16–17; day, 18–19; night, 18–19; normal, 131; spatial, 12, 15–17, 24–7
Diversity index, 22–4; of species, 21–4, 68
Dock, 259, 262
Dodder, 39, 241
Dogwood, 250
Domesday Book, 249
Donkey, 4
Downland communities, 166–71; chalk, 166; climax vegetation, 167; grazing pressure, 167; limestone, 166; succession, 167–8

Dragonfly, 186, 245
Dronefly, 187
Duckweed, 190, 205
Duffey, E., 249, 290
Dugesia gonocephala (flatworm), 201
Dune: animals, 178; 'blow-out', 178; environment, 175–7; fixed (grey), 174; grassland, 175; heath, 175, 176; marsh, 175; partially fixed (yellow), 173; rabbits in, 178; slacks, 175; succession, 172–5; young, 172
Dutch elm disease, 33–4, 49
Dwarf willow, 175, 245
Dyer's rocket, 256
Dysdera erythrina (spider), 258

Earthworms, sampling, 81–2
Ecdyonurus venosus (mayfly), 203
Echo sounding, 64
Ecological genetics, 9, 26, 144
Ecological technology, 284–6
Economics, 1, 2
Ecosystem, 3, 57–8, 149; modification of, 278–80
Ectoparasite, 38
Edaphic factor, 1, 16, 98
Edington, J. M., 19
Edlin, H. L., 276
Eel grass, 232
Eisen's worm, 245
Elder, 175, 250, 256
Elephants, in game reserves, 65–6
Elm bark beetle, 34
Elton, C., 1, 12, 31, 54
Emigration, 28
Emperor moth, 243–4
Encarsia formosa (wasp), 44
Endoparasite, 39, 45
Energy conversion, 51; flow, 57–8
Enteromorpha intestinalis (alga), 220, 232
Ephemera danica (mayfly), 205
Ephemeral, 175, 259
Ephemerella sp. (mayfly), 23, 205
Ephydatia fluviatilis (sponge), 202
Epibiont, 54
Epiphyte, 54
Epizoite, 54
Erosion, 19–20
Estuary, 227–34; environment, 227–8; life in, 228–31
Eupagurus bernhardus (crab), 40
Euryhaline species, 229
Eurythermous species, 184, 201, 213
Eutrophic conditions, 199
Evaporation, latent heat of, 50
Evolution, 1, 9
Extinction, 12, 289
Extinction point, 160
Eyebright, 241

Facultative parasite, 39
Feather moss, 251
Feeding habits, changes in, 284
Fertility, 40
Festuca spp. (grass), 24–6, 259
Field maple, 250
Field mouse, 243, 251
Field vole, 34, 243, 251
Fireweed, *see* rosebay willowherb
Flatworm (Cestoda), 12
Fletcher, W. W., 44
Flotation, 82
Flounder, 229
Flour beetle, 14–15, 90
Flowvane, 110
Fly agaric, 41, 241
Food and Agriculture Organisation (FAO), 47, 282
Food, as an ecological factor, 37–40, 47
Food chain, 3
Food web, 51–4, 57, 62–6
Fool's watercress, 191
Forestry Commission, 277
Ford, E. B., 274
Foxglove, 292
Fox moth, 243
Frequency, percentage, 93
Friedrich, K., 31
Fruit fly, 30
Fry, G. L. A., 23
Fungicide, systemic, 287

Game reserve, management, 65–6
Gammarus spp. (crustacean), 53, 54, 86, 136–8, 183, 189, 202, 230
Garden snail, 251, 257
Gatekeeper butterfly, 251
Geiger counter, 37
Gene pool, 70
Genetic interest, in ecology, 291
Gills, 186
Gipsy moth, 278–80
Glasswort, 228, 232
Glow-worm, 168
Gnat, 186
Golden samphire, 219
Gorse, 239
Gradwell, G. R., 34
Graph, 127–8
Grasshopper, 251, 277
Gravestones, colonisation by lichens, 269–70
Grazing pressure, 167
Great water grass, 196
Greenfinch, 288
Grey scalloped bar moth, 243
Ground elder, 263
Groundsel, 7, 259, 263

Habitat, 12, 62, 149; preferences, 14–15
Haeckel, E., 1
Haemoglobin, 185
Hairy screw cress, 15, 263

Halophyte, 172, 219, 228
Halosere, 232
Hardy, Sir A., 64–5
Harper, J. L., 29
Hart's tongue fern, 251
Hawthorn, 168, 239, 250, 288
Hazel, 156, 250
Heath, O. V. S., 85
Heath community, 239–43; formation, 236–7; succession, 246
Heath rush, 247
Heavy metal tolerance, 24
Hedge bedstraw, 251
Hedge brown butterfly, 170
Hedgerow, 249–52; as a dynamic system, 251–2; communities, 250–1; destruction, 275
Hedge sparrow, 271
Helicella itala (snail), 168
Henbane, 292
Hepburn, I., 175
Herbicides, effects, 262, 275
Hermaphroditism, 40
Hermit crab, 40
Herring, 64
Heterotrophs, 50, 68
Histogram, 129–30
Holcus lanatus (grass), 251
Holist approach, 3, 6
Holly, 156
Holocoenotic community, 31
Homeostasis, 49
Honey fungus, 158
Hornwort, 184, 192
Horse, 4
Horse leech, 245
Horseshoe vetch, 167
Howard, E., 37
Hudson Bay Company, 31
Humidity, as an ecological factor, 32, 38, 108–9, 214–16
Humphreys, T. J., 232
Humus, in soil, 102–3
Hydrogen ion: concentration as an ecological factor, 112, 188–90; determination, 112
Hydrological cycle, 58–9
Hydropsyche siltalai (caddis fly), 19–20
Hydrosere, 15, 30, 89, 196, 205
Hygrometer, 108

Ichneumon, 37, 42–3
Illumination, 107
Immigration, 28
Immunity, 45
Indicator species, 21, 122
Industrial revolution, 268
Insectivorous habit, 241
Integrated control of pests, 45
Interactions between species, 40
Intermediate density-dependent factors, 31
International Society of Soil Science, 100

Iron bacteria, 62
Irradiance, 107
Ivy, 251
Ivy-leaved toadflax, 256

Jenkins's spire snail, 23, 188, 204

Kestrel, 243
Kettlewell, H. B. D., 36, 271–4
Knotted wrack, 216–19, 220–1

Lady's fingers, 167
Lake Windermere, 19
Lapwing, 275
Larch, 276
Large blue butterfly, 290
Large white butterfly, 42–3
Laundon, J. R., 269
Leadley Brown, A., 85
Lecanora spp. (lichen), 254, 269, 271, 273
Leech, 245
Lemming, 30
Lepraria sp. (lichen), 271
Lichen, 41, 254
Light, as an ecological factor, 7, 8, 38, 57, 106–8, 159–60, 182–4, 212–13
Light meter, calibration, 107; integrating, 57, 108; photographic, 107; uses, 108
Limpet, 214
Lincoln Index, 94–5
Ling (fish), 28
Ling (heather), 21, 236
Linnaeus, K., 6
Lithobius forficatus (centipede), 258
Litter layer, in soil, 153
Littorina spp. (periwinkle), 21, 213, 214, 216, 220–2
Liver fluke, 39
Longworth mammal trap, 78
Lousewort, 239, 241, 243
Lugworm, 214, 224, 231, 233
Lumbricus rubellus (earthworm), 81
Lumbricus terrestris (earthworm), 81, 149
Lumpers, 5
Lungs, 188
Lynx, 31

Macan, T. T., 16, 189
Macroclimate, 109
Maidenhair spleenwort, 256
Malaria, 39, 285
Malathion, 9, 45
Man, as an ecological factor, 2, 9, 268–92
Management, ecological, 144, 288
Marbled white butterfly, 171
Margalef, R., 69
Mark-release-recapture, estimation of numbers, 34–5, 94–6
Marking methods (animals), 90–1

Marram grass, 173, 176
Marsh fritillary butterfly, 32
Mayr, E., 5
McNeilly, T. S., 25
Meadow brown butterfly, 17, 37, 70, 171, 250
Meadow pipit, 243, 278
Meadowsweet, 158, 252
Mean, 85, 130–1
Melanic moths, 271–4
Melanism, industrial, 24, 271–4
Mellanby, K., 286
Methanal, 81
Microclimate, 109
Mildew, on cereals, 286–8
Milkwort, 166, 168, 169, 239, 240
Minerals, as an ecological factor, 8, 190, 241
Minnow, 187, 192
Monitors, multi-purpose electronic, 117
Moon, H. P., 16, 189
Moorland, 238
Mor, 102, 151, 237
Morphology, 1
Mosquito, 39, 285
Motorway, salinity, 233–4
Mottled beauty moth, 274
Mowing, effects, 259
Mud, 192–3, 205, 231
Mull, 149, 237
Mutualism, 40, 70
Mycorrhiza, 158, 241
Myxomatosis, 44, 49, 168

Natural selection, 70
Nature Conservancy, 291
Nature conservation, 288–92
Neal, E. G., 51, 53
Neap tide, 210
Nelson-Smith, W., 120–1
Nematode (roundworm), 39
Net, drag, 76; entomological, 72–3; general collecting, 73; plankton, 73; sweep, 72–3
Neutralism, 40
Newell, P. F., 18
New Forest, 292
Niche, ecological, 12, 62, 66, 68, 161, 162, 224–5
Nicol, J. A. C., 216
Nitrate deficiency, 241
Nitrification, 61
Nitrogen cycle, 53; fixation, 61
Nodules, of leguminous plants, 61
Normal distribution, 131–2
Null hypothesis, 139
Numbers, balance of, 49–50; fluctuation, 168
Nuthatch, 271

Oak, 168, 250
Oak eggar moth, 243
Oak-leafroller moth, 32
Oak mildew, 34
Odum, E., 30, 40, 46
Operculates, 188, 214

Operculum, 188
Osmotic control, 229–31
Oxford ragwort, 233, 256
Oxygen, as an ecological factor, 7, 113–17, 185–8; estimation, 114–17, 124–6; in sand and mud, 117; sampling, 113–14
Oyster, 231

Pan, 151, 237
Papilionaceae, 7
Parasite chain, 54
Parasitism, 2, 12, 38–46, 243
Parmelia saxatilis (lichen), 269
Partridge, 275
Paths, 261–2
Pearlwort, 259
Pearsall, W. H., 238
Pea-shell cockle, 183, 193
Peat, 237
Peppered moth, 5, 24, 271–4
Percentage cover, 88
Periwinkle, *see Littorina* spp.
Perring, F. H., 286, 288
Permethrin, 45
Pesticide, 44–5
pH, *see* hydrogen ion concentration
Phillipson, J., 51, 85
Phosphorus, 62
Photoconductor, 107
Photometer, integrating, 108
Photosynthesis, 7, 12, 183
Photovoltaic cell, 107
Physical factors, 2, 7, 15–20, 32, 38
Phytoseiulus persimilis (mite), 44
Pineapple weed, 259, 263
Piptoperus betulinus (bracket fungus), 41
Pitot tube, 111
Plankton, 62–5, 195–6
Plankton recorder, 64–5
Plantain, 259
Playing fields, ecology of, 259–63
Plasmodium spp. (malaria parasite), 39, 285
Platyerthus hoffmansegii (woodlouse), 40
Plectronemia conspersa (caddis fly), 19–20
Plume moth, 243
Podsol, 151, 237
Point frame, 88–9
Pollution, 2, 24–7, 268–75
Polycelis cornuta (flatworm), 23
Polycelis nigra (flatworm), 189, 202
Polycentropus sp. (caddis fly), 202
Polygenic inheritance, 5
Polymorphism, genetic, 5
Ponds, 181–99; communities, 190–6; environment, 181–90
Pondskater, 191
Pond snail, 53, 188
Pondweed, 53, 184

Pooter, 77
Population density, 28, 36; dynamics, 144; estimation, 91–6, 143; fluctuation (animals), 34–8; fluctuation (plants), 32–4; growth, 29–31, 144; human, 46; size, 28–40
Portland, Dorset, 9
Potometer, 108
Prawn, 233
Predator, 40, 41, 44–5
Predator–prey relationships, 144
Prediction, in ecology, 143–4
Presentation of information, 127–30
Preservation, of the environment, 288
Primrose, 5, 139–42, 157
Privet, 156, 175
Probability, 133
Producer, primary, 9, 50–1, 69
Production, gross, 56; net, 56, 68
Productivity, 56
Profile, chart, 89
Profile, vertical, 82, 100, 150
Propagule, 7, 21, 28, 252
Protein, supply, 51, 282–4; synthesis, 284
Protocooperation, 40
Pteromalus puparum (wasp), 36
Pulmonate, 188, 214
Purple moor grass, 239, 241
Pyramid, ecological, 54–6; of biomass, 55; of energy, 57; of numbers, 54–5

Quadrat, 86–9; edge effect, 88; permanent, 86–7

Rabbit, 44
Ragworm, 231, 233
Ragwort, 10, 30, 262
Rainbow trout, 201
Rampion, 44
Rape, 276
Ray, J., 5
Recorder, multi-channel, 117–18
Recycling, of resources, 281–2
Red ant, 40
Red deadnettle, 251
Red fox, 31
Red spider mite, 44
Red squirrel, 277
Red valerian, 257
Reductionist approach, 3, 6
Resistant strain, to pesticide, 9, 275
Reproduction, 12
Respiration, 7, 50, 68; pigment, 185
Restharrow, 275
Reynoldson, T. B., 16, 189
Rheotaxis, 203
Rhizobium sp. (bacterium), 41, 61, 241

Ribwort plantain, 25
Richardson, D. H. S., 268
Ringing, birds, 90
Ringlet butterfly, 170
River lamprey, 39
River Thames, 58, 274
Roach, 192
Robin, 278
Rock rose, 168
Rosebay willowherb, 7, 10, 28, 30, 158, 159, 266
Rosette growth, 168, 260
Royal Society for the Protection of Birds, 290
Rye grass, 259

Sacculina sp. (barnacle), 39
Salisbury, Sir E., 168, 175, 177, 262
Salvage, 280
Samphire, 219
Sampling, crude procedures, 85–6; error, 86, 131; methods, 85–91
Sand goby, 213
Sand sedge, 173
Saprophyte, 12
Sawfly, 245
Scale insect, 45
Scalloped hazel moth, 271
Scarlet tiger moth, 34, 37
Scott, N. E., 233
Scot's pine, 239, 276
Sea anemone, 40, 214
Sea aster, 228, 232, 233
Sea buckthorn, 174, 175
Sea holly, 172
Sea lavender, 232
Sea lettuce, 232
Sea meadowgrass, 234
Sea plantain, 233
Sea purslane, 228, 232
Sea rush, 232
Sea spurge, 172
Sea spurry, 233
Seashores, 210–27; environment, 210–16; habitats, 224–6; zonation, 210–12, 216–24
Seawater, composition, 118
Seaweeds, 21, 214, 216–19, 220, 221, 232
Sea wormwood, 178
Secchi disc, 106, 182
Secondary host, 39
Selective advantage, 9
Self-fertilisation, 40
Self-heal, 167, 259, 260, 263
Semiconductor diode, 105
Sequential comparison index, 23
Sere, 67
Serrated wrack, 216–19
Shade plants, 159
Sheep's fescue, 166
Shepherd's purse, 259, 263
Shock disease, 34
Shore crab, 39, 229
Silver birch, 41
Silverweed, 263

Simocephalus sp. (crustacean), 18
Simulation, 69, 143
Simulium sp. (midge), 202
Site of Special Scientific Interest (SSSI), 291
Six-spot burnet moth, 171
Skylark, 275, 278
Slug, 18
Small blue butterfly, 168
Small elephant hawk moth, 10, 266
Small heath butterfly, 243, 277
Small tortoiseshell butterfly, 266
Snow-shoe rabbit, 31
Soil, air spaces, 151; auger, 99–100; burrowers in, 81, 152; classification, 100–3; communities, 151–3; composition, 98–103; mechanical analysis, 100–2; organisms, extraction, 81–5; sampling, 82, 99; water, 102, 151; zonation, 100, 150
Song thrush, 168, 259
Soot, 273
Southern, H. N., 38
Southward, A. J., 221
Specialisation, 39
Species, concept, 1, 3–5, 6; criteria, 4–5; diversity, 21–4, 68; polytypic, 5
Speckled wood butterfly, 171
Speedwell, 251
Spellerberg, I. F., 290
Sphaeroma sp. (crustacean), 230, 233
Spider crab, 229
Splitters, 5
Spotted flycatcher, 271
Spotted orchid, 168, 169
Springtails, 191
Spring tide, 210
Spruce, 276
Square-tailed worm, 185
Squirrel, 277
Sri Lanka, malaria in, 286
Standard deviation, 85, 87, 131–2
Standard error, 87, 132–4
Standing crop, 55, 66
Starwort, 184, 192, 203
Statistical methods, 131–42, 143
Statistical significance, 133–4
Steel production, 280
Stenohaline species, 229
Stenothermous species, 184, 201
Sterilisation, 46
Stickleback, 192
Stinging nettle, 266
Stonecrop, 175, 240, 254
Stone loach, 205
Stream, 199–207; environment, 201–5; succession, 205–7
Stress syndrome, 34, 47

Succession, 10, 67–70, 196–8, 231–3; evolutionary implications, 70; time factor, 68–9
Sulphur-35, 36
Sulphur bacteria, 61–2
Sulphur dioxide, 269, 273
Sundew, 236, 241, 243, 246
Sun plant, 7, 159
Surveying methods, 120–2
Swan mussel, 1, 93
Symbiosis, 2, 61
Symbiotic bacteria, *see Rhizobium*
Synthesis, 3
System, 3

t-distribution, 134–5, 146
Tabulaton, 127
Tagging, fish, 90
Tanypus sp. (midge), 188, 245
Tapeworm, 39
Tawny owl, 38
Taxis, 183
Taxonomy, 1, 3
Temperature, as an ecological factor, 7, 103–5, 184–5, 213
Territorial behaviour, 37–8, 66–7, 91
Territory, 37–8
Thermistor, 104
Thermocline, 184
Thermocouple, 105
Thermodynamics, First Law of, 51
Thermometers, 104–5
Thiabendazole, 287
Thigmotaxis, 203
Three-spined stickleback, 192
Toadrush, 245
Top shell, 220
Tor grass, 166, 168
Tormentil, 239
Tortella tortuosa (moss), 255
Tortula ruraliformis (moss), 174, 176
Tourism, 292
Trampling, effect, 260
Transect, belt, 89; line, 89, 93
Tranter, J., 117, 118
Traps, light, 79–80; mammal, 78–9; pitfall, 78; water, 79
Tree pipit, 278
Trophic levels, 10, 51, 54, 57
Trout, 7, 8
Trumpet snail, 53
Tubifex sp. (midge), 185, 193
Tullgren funnel, 82–4, 265
Turrill, W. B., 141, 142
Tyrosine, 271

Uric acid, 216
Usnea spp. (lichen), 268, 271

Valerian, 158
Variance, 85; analysis of, 136
Variates, 85
Variation, 9
Varley, G. C., 34
Varley, M. E., 201
Vector, 39, 285

Vinegar eelworm, 29
Vision, 183
Vole, 251

Wall: animals, 256–9; environments, 254–6; plants, 254–5; succession, 256
Wall pepper, 167, 254
Wall speedwell, 256
Wandering snail, 188, 202
Warfarin, 9
Waste-derived fuel (WDF), 280
Wasteland populations, 263–6
Water boatman, 187, 247
Water, buoyancy, 181; capillary, 102; chemical composition, 118, 188; current, 201–4; geological factors, 240–1; gravitational, 102; hardness, 119; hygroscopic, 102; in soil, 102–3; osmotic, 102
Water crowfoot, 192, 202, 205
Water cricket, 191
Water dock, 196
Water fern, 190
Water lily, 191
Water louse, 16, 119, 183, 189
Water milfoil, 205
Water mint, 191
Water scorpion, 186
Water snail, 39
Water starwort, *see* starwort
Weed killers, 262
Westlake, D. F., 108
Wheatear, 243
Wheatstone Bridge, 104
White, R. W., 150
Whitefly, 44
White-lipped snail, 9, 130, 168, 251, 257
Widger, 80
Willow moss, 202, 205
Willow, 175, 263
Winkler method, oxygen estimation, 115, 124–6
Wood anemone, 157
Woodlouse, 40, 258
Wood-Robinson, C., 121
Woods, classification, 154–5; communities, 154–62; layers, 155–8; stratification, 155–8
World Health Organisation, 285
Wormery, 82
Wormwood, 178
Wratten, S. F., 23
Wytham Wood, Oxford, 12, 34, 38

Xanthoria sp. (lichen), 15
Xerophyte, 172, 219
Xerophytic conditions, adaptations for, 240–1
Xerosere, 15, 16

Yapp, W. B., 92, 160, 275, 277
Yellow ant, 40
Yew, 256

Zooxanthellae, of corals, 41